Eduardo Santos
Pedro Lúcio Barboza
Josemberto R. da Costa

Spatial Geometry

Eduardo Santos
Pedro Lúcio Barboza
Josemberto R. da Costa

Spatial Geometry

Teaching and Learning

ScienciaScripts

Imprint
Any brand names and product names mentioned in this book are subject to trademark, brand or patent protection and are trademarks or registered trademarks of their respective holders. The use of brand names, product names, common names, trade names, product descriptions etc. even without a particular marking in this work is in no way to be construed to mean that such names may be regarded as unrestricted in respect of trademark and brand protection legislation and could thus be used by anyone.

Cover image: www.ingimage.com

This book is a translation from the original published under ISBN 978-3-330-99569-7.

Publisher:
Sciencia Scripts
is a trademark of
Dodo Books Indian Ocean Ltd. and OmniScriptum S.R.L publishing group

120 High Road, East Finchley, London, N2 9ED, United Kingdom
Str. Armeneasca 28/1, office 1, Chisinau MD-2012, Republic of Moldova, Europe
Managing Directors: Ieva Konstantinova, Victoria Ursu
info@omniscriptum.com

Printed at: see last page
ISBN: 978-620-8-58178-7

The only way to learn maths is to do maths.

Paul Halmos

SUMMARY

INTRODUCTION

Considering the changes in the curriculum of primary and secondary schools, the need for schools to keep abreast of technological advances in our environment has become increasingly evident. New technologies are increasingly present in our daily lives, from the personal to the professional.

Based on this, we believe it is of great importance for teachers, as professional mediators between school and society, to review their work proposals, integrating technological resources into the curriculum in order to make a significant contribution to learning.

In the case of maths, there are many difficulties students have in understanding certain content. If we look specifically at the study of Geometry, and even more specifically Spatial Geometry, these difficulties are manifested by the vast majority of students. In view of these circumstances, interest arose in a study involving the use of computer resources in the teaching of Spatial Geometry, since the school in which the research was carried out has a computer laboratory, even though it has few computers in good working order. The decision to use the GeoGebra 3D software was based on the criterion that it is a well-designed didactic application that allows us to construct three-dimensional figures in a simple and dynamic way.

Given this context, this study aims to analyse the contributions of the GeoGebra 3D software in the teaching of Spatial Geometry in a state school in the municipality of Juarez Távora.

In order to achieve these objectives, we have set ourselves the following specific goals:

C Build and plan geometric solids;

D Determine the areas and volumes of the main polyhedra;

C Build solids of revolution using the software;

R Relate geometric solids to everyday objects;

C Understand the importance of using new technologies to teach maths;

A Analyse the results obtained using the GeoGebra 3D software.

We refer to authors such as Valente (1999), Moysés (1997), Candau (1991), Borba (2007) and others who discuss the importance of using technology to teach maths, because like them we understand that it is through new ways of teaching that we can achieve a quality education that takes into account the interests and needs of the students.

Methodologically, we chose to carry out this study using a descriptive and exploratory qualitative approach, which allows us to analyse the data in depth. We used a questionnaire as a data collection tool. The study's investigative field was a secondary school class in a public school. The research subjects were 32 students from the 2nd year of secondary school, in the morning shift, and a maths teacher from the same institution.

The work is structured in four chapters. The first chapter deals with issues relating to the process of teaching and learning maths. We consider the difficulties that students face in learning and that teachers face in teaching. In the second chapter, we present **the theme "The** Use of Technology in Education", addressing the contribution of technology to learning and the myths surrounding its use as a pedagogical tool. The third chapter, **entitled "Knowing the Methodological Path of the Research"** presents the methodological procedures, the research site, the research subjects, the instruments used and the entire data collection process.

In the fourth chapter, we analyse and discuss the data obtained. Describing the results of using the software as a teaching tool. In this chapter we will also present the opinions of the research subjects, obtained through the data collection instrument of questionnaires, seeking information on the use of the GeoGebra 3D software and the factors that interfere with learning maths.

Finally, we present our final thoughts on the subject and point out the relevance of this work, as it is hoped to stimulate new studies and provoke questions about the use of GeoGebra 3D software as a pedagogical tool that helps in the teaching process, contributing significantly to the development of student learning.

CHAPTER 1

THE MATHS TEACHING AND LEARNING PROCESS

Throughout the evolution of humanity, maths has made important contributions to countless areas of knowledge. In a brief look back at history, we can see, for example, the importance of arithmetic in the expansion of trade and in interest calculations with the emergence of banking institutions. In another branch of knowledge, for example, the progress of astronomy would have been impossible without the presence of maths.

If we take a closer look at the last few decades, we can see the exponential growth of technology with mathematical foundations as one of its main tools. In short, maths is present in every situation. If we look around us, we can easily see its presence in contours, in the shapes of objects, in measurements, in domestic, school and urban activities.

However, over time, maths has become a real "bogeyman" for the majority of students who encounter this subject. Some factors have contributed to worsening this situation, such as poorly prepared teachers, outdated teaching resources, schools with poor infrastructure, the use of ineffective methods in the classroom and a lack of student interest.

Somehow, maths teachers are perplexed when they say that they try so hard, that they explain the content very well, but the students don't learn. They almost always say that students don't learn because they don't study, they don't have the commitment to study or because they don't want to learn.

Teachers haven't realised what obstacles and difficulties need to be overcome in teaching or learning, which need to be tackled in the classroom and beyond. It doesn't take much effort to realise that the story told by teachers, blaming students for their lack of interest or failure, doesn't fit reality, because there are a number of elements that contribute to the poor learning of children and young people in classrooms at all levels of education.

In recent decades, the way in which school content has been worked on has been criticised by many researchers. For Moysés (1997, p.59), **"it is as if the process of schooling encouraged the idea that in the 'school game' what counts is** learning various types of symbolic rules, learning that must be demonstrated **within it". This methodology** has caused much of the student body to reject this subject which, as we have seen, has played a fundamental role in historical and social progress.

According to Silveira (2002), students' opinions about maths reveal repeated meanings from other voices, whether from teachers or even other students. The author also reveals that students' dissatisfaction is expressed by "Maths is boring", "I don't like maths", "maths is difficult". This last opinion was impregnated into the subject of mathematics and came to be recognised not only by the students, but also in the historical context of the subject. **This "fame" that has given voice to teachers and** students demonstrates the naturalised and unquestionable way in which mathematical knowledge is constituted at school: mathematics is traditionally the subject that presents the greatest difficulty. Thus, we can perceive the discourse that speaks of the difficulty of maths as a pre-constructed discourse (SILVEIRA, 2002).

It is well known that maths in ancient times arose out of man's need for everyday life. However, with the influence of the movement known as Modern Mathematics, schools have prioritised teaching that focuses mainly on abstractions with excessive use of formulas and symbols that most of the time have no meaning for the student.

Many researchers have expressed concern about this teaching method, which for some is already seen as a failure. But what would meaningful maths learning be? According to Madruga (1996, p.43),

> Meaningful learning is distinguished from other types of learning by two characteristics: the first is that its content can be related to the student's previous knowledge, and the second is that the student must adopt a favourable attitude towards this task, giving the content they assimilate its own meaning.

From this perspective, the teacher's task is to programme, organise and sequence the content so that students can carry out this learning, incorporating new knowledge into their previous cognitive structure. In a similar vein to Madruga, Moysés reports on the school's failure to take into account prior knowledge acquired by students outside the classroom:

> [...] one also realises that knowledge acquired outside of school is not always used as a basis for school learning. What's more, it's not even taken into account as a motivational resource (MOYSÉS, 1997, p. 60).

This thinking comes not only from researchers concerned with this teaching method, but can also be identified in the National Curriculum Parameters (PCN), which offer elements that broaden the national debate on the teaching of this area of knowledge.

> The importance of taking students' prior knowledge into account when constructing meaning is also often overlooked. Most of the time, the concepts developed in the course of the child's practical activity, of their immediate social interactions, are underestimated, and the school treatment is started in a schematic way, depriving the students of the wealth of content coming from personal experience (BRASIL, 1998, p.23).

School maths has conveyed the idea of an "isolated science", where numbers, calculations, rules and many other elements seem unrelated to the world around them. This requires students to have knowledge that is of little use or significance to their daily needs, often forcing them to acquire knowledge that will be of little use to them outside the four walls of the classroom. How many teachers often hear classic questions like: What is this for, teacher? Where will I use this, teacher?

Unfortunately, most of the time the teacher is unable to give the student a convincing and satisfactory answer. The answers to these questions are always very similar, such as: Study because it's for the entrance exam; if you're going to study engineering, architecture or something related, you'll need it. In reality, these answers serve more to get the teacher out of a jam than to satisfy the student's curiosity. It's not hard to understand why teachers don't find the answer that the student really wants. After all, they already teach what they've learnt, and it's quite likely that the maths they've been taught has also been disconnected from reality. Thus maintaining a traditionalist model in maths teaching.

Traditional teaching is characterised by its emphasis on abstract, formal, mechanised, expository and decontextualised mathematics, most of the time excluding the student's socio-cultural intersections. This teaching model emphasises the teacher as the "master of knowledge", whose role is to transmit knowledge to the student, who plays the role of mere receiver. In practice, traditional teaching treats the student as an individual who knows nothing, who is there to learn and reproduce everything exactly as they supposedly learnt it.In today's world, faced with so many technologies and changes in the social environment, it is not possible to continue being that teacher of the past, with the same methodologies and attitudes towards the teaching-learning process. The National Curriculum Parameters clearly and succinctly characterise this practice, which is already considered to have failed, but is still widely used in classrooms:

> Traditionally, the most common practice in maths teaching has been one in which the teacher presents the content orally, starting with definitions, examples, demonstrating properties, followed by learning, fixing and application exercises, and assumes that the student learns by reproduction. Thus, correct reproduction is considered to be evidence that learning has taken place (BRASIL, 1998, p.37).

This current situation has caused student demotivation and, consequently, learning difficulties. Changing this reality is necessary so that a new relationship between teachers and students can begin to exist within the school.

The teacher-student relationship, in turn, has also sparked various discussions in the teaching and learning process. According to Sousa Júnior and Barboza (2013, p. 208),

> Teacher-student interaction is a factor that contributes to and influences student learning, as well as the student-student relationship, in which one colleague helps another by discussing and formulating hypotheses about a problem, contributing to their cognitive development and thus developing their mathematical reasoning. Sometimes the teacher doesn't know what their reasoning and understanding of maths really is.

The way the class is dealt with has a direct impact on the acceptance or rejection of the subject. If students aren't interested in the subject, either in the teacher or in the lessons, they find it very difficult to learn, and this discourages them. Their disinterest and aversion to the subject, and even to the teacher, increases. A vicious cycle of demotivation and failure to learn sets in, which is difficult to break. In addition to their theoretical and practical knowledge of teaching techniques and methods, teachers need to be aware of aspects and conditions that motivate students in the classroom, making teaching and learning more enjoyable and stimulating.

Demotivation negatively affects the teaching-learning process. Students often lose motivation when their basic needs are not met. Nowadays, it has been a tremendous challenge for teachers to keep students motivated and wanting to learn. One thing no longer seems to be up for discussion - something needs to change. Even the most traditional and conservative teachers admit that they can no longer limit themselves to chalk and blackboard. The search for innovative alternatives in maths teaching has become extremely necessary.

According to Lorenzato (2006), the teacher plays a fundamental role in both the success and failure of students at school. For this author, the teacher having good teaching material is not enough to guarantee meaningful learning.

Of all the difficulties faced, we would highlight the working conditions offered by the public authorities, with insignificant salaries paid to teachers, and the application of insufficient resources to education, not to mention the responsibility of governments at all levels of the federation.

On the other hand, we have problems that involve more than one dimension, that is, in addition to public authorities, other actors are jointly responsible for the failure of learning. The methodology used in the classroom does not contribute to a change in reality. This methodology, which doesn't seem to be appropriate for the needs of today's children and young people, is the result of a teacher who has graduated from a degree course and has failed to keep up with the changes and transformations that have taken place in society.

It turns out that the difficulties faced by students in maths are not being perceived by teachers. Teachers somehow find it difficult to put into practice the proposals that have been put forward, for example in what the PCN proposes: "Learning in maths is linked to understanding, that is, to grasping meaning; grasping the meaning of an object or event presupposes seeing it in its relations with other objects and events (BRASIL, 1998, p. 19). This pedagogical approach suggested by the PCN still seems far removed from maths teachers' practice. Teaching without meaning results in students not learning.

Soistak, Pinheiro and Pilatti (2011) state,

> It is known that learning does not only occur when content is presented in an organised way, nor even when students repeat models they have studied. Learning with understanding is more than giving the right answers, it's preparing to face new situations, establishing links between the new and the known, it's knowing how to create and transform what you already know (SOISTAK, PINHEIRO, PILATTI, 2011, p. 38).

You can only learn with understanding if the reciprocal is also true for teaching, that is, if teaching is carried out with understanding for the student. And what we want from teachers is for them to endeavour in their daily classroom practice to make teaching and learning with understanding possible. This is a goal that still takes some time to achieve and requires desire and effort on the part of both public authorities and teachers.

Another relevant issue is emphasised by Vygotsky when he states that the teacher's task is to develop not a single capacity for thinking, "but many particular capacities for thinking in different fields; not to strengthen our general capacity for paying attention, but to develop different faculties for concentrating attention on different subjects" (VYGOTSKY, 2006, p. 108). In other words, for the author, each capacity must be developed independently through appropriate activity.

In the 21st century, we still see the practice of traditional teaching. Traditional maths teaching is characterised by certain ways of organising the classroom. According to Alro and Skvsmose, in this model, "lessons are usually divided into two parts: first the teacher presents some mathematical ideas and techniques, usually in accordance with a textbook. Then the students do some exercises by directly applying the techniques presented" (ALRO; SKVSMOSE, 2006, p. 51). The way in which the authors present traditional teaching is somewhat contested by Sousa Júnior and Barboza (2013), who claim that the teacher does something even more uninteresting in the classroom. The teacher just explains the content on the chalkboard and the student copies it, in which case the student doesn't do any exercises to apply the technique. The exercises are solved by the teacher.

In maths teaching there is a very common practice of teaching standard answers to standard questions. Alternatives to traditional teaching should be sought. One practice that can be incorporated is problem solving, in which the student will not need standardised techniques and formulas to answer, but will have an attitude of investigation towards what has been proposed, having to activate knowledge in order to overcome obstacles to solve the problem. Situations in which the student needs to try out various solutions and create hypotheses to solve the proposed problem, developing mathematical reasoning. For this to happen, the teacher is very important, as they need to know how to guide and stimulate the student, proposing questions that make it possible to develop understanding and are linked to everyday life, so that they can feel challenged by the proposed question.

Activities with manipulative teaching materials can also be part of the resources used by teachers, as they are an important teaching resource. As well as being a material that makes lessons more lively, dynamic and attractive, they make it possible to relate mathematical theory to practice.

School is the place where intentional pedagogical intervention triggers the teaching and learning process, and it is the teacher's role to intervene in the process, unlike informal situations in which children learn through immersion in a cultural environment. One point that needs to be considered is the fact that children's learning begins before school learning (VYGOTSKY, 2006). A child's learning does not begin at school. When children go to school to study arithmetic, they already bring with them the experience they have acquired about quantity in everyday life. In fact, the child has already undergone their own learning process, which needs to be respected and taken into account. That's why it's essential to consider the diversity that exists in the classroom, so that everyone is included and has the right to participate and transform the social context according to their needs.

Vygotsky **says that "what the child can do today with the help of adults, he will be able to** do **tomorrow on his own"** (VYGOTSKY, 2006, p. 113). This statement should help teachers understand the important role they play in the classroom, which is not just to transmit content or make copies on the blackboard for students to reproduce. The teacher guides the student, helps only when necessary, interferes when appropriate, always suggests, questions or inquires a lot, reflects on knowledge and reality, provokes restlessness and curiosity in the student.

Currently, a lot of research in the field of maths has emphasised the development of innovative methodological alternatives. Maths education is being considered beyond the blackboard. Among the many possibilities and methodologies proposed, one alternative that has been emphasised and which can greatly contribute to the teaching-learning process in this subject is the use of new technologies in the classroom. According to Fernandes, "in an increasingly globalised world, using new technologies in a way that is integrated into the pedagogical project is a way of getting closer to the generation that is on the **school benches"** (FERNANDES, 2010, p. 46). It's obvious that the mere fact of including technological resources in maths teaching doesn't mean learning. Quality is needed in their use, and this quality will depend on how the proposals are interpreted by the teachers.

It's common to see in the school environment a certain amount of fear on the part of teachers when it comes to including new technological resources in education. This fear is often transformed into a feeling of insecurity or even resistance to changing teaching practice, as teachers in this new position are challenged to review and expand their knowledge in order to deal with new situations. Faced with such circumstances, adequate teacher training becomes indispensable. The teaching of maths requires new methodological strategies and, for this, the teacher needs to go in search of and be qualified for new didactic experiences, because otherwise, a lesson using technological resources can be just as traditional as one conducted only with a blackboard and chalk, if the teacher is not sufficiently prepared and convinced that the use of new technologies can be an ally in the process of teaching and learning maths. It's important to emphasise that the aim of integrating technological resources is not to replace classical work in the subject. Practices such as mental calculation and sketching graphs with a pencil and eraser are still essential for developing mathematical reasoning. It is also important to remember that the most legitimate and powerful "resources" for learning remain the teacher and the student who, together and interactively, will be able to find new ways of constructing knowledge. Technological resources have the role of integrating this composition as a tool in the construction of knowledge. In this way, the appropriate use of technology is an aid to the teaching-learning process.

CHAPTER 2

THE USE OF TECHNOLOGY IN EDUCATION

It is unlikely that teaching practices will follow the models of previous centuries, limited to four walls, restricted in time to class hours and based on a relationship in which someone who has the knowledge passes it on to others. The transformations underway are likely to significantly change teaching practices and knowledge production processes.

In today's world, it's almost impossible to talk about education without referring to technology. Because schools, like many other segments of society, are permeated by various types of technology.

Technology is present in Brazilian schools, making it necessary to think about how to use it to contribute to the teaching and learning process. A lot of research (BITTAR, 2010; BORBA and VILLARREAL, 2005; MARIN, 2012; VALENTE, 1999; KAWASAKI, 2008; MARTINI and BUENO, 2014; CARNEIRO and PASSOS, 2009; COAN et al., 2013) has been dedicated to investigating this issue and points to the contributions of using technology to teach maths.

Some of these studies point out that the use of information and communication technologies (ICT) in the classroom can contribute to greater learning, the stimulation of creativity, the possibility of testing conjectures and modelling mathematical problems, as well as many other activities that would be unfeasible in a paper and pencil environment, because they require a lot of time or deal with precision errors that would negatively influence the objective of an activity.

Over the years, the implementation of technology has brought society various changes and facilities that have been quickly and significantly accepted, understood and improved. Through technology, human beings have gained agility, comfort and efficiency in areas such as communication, domestic activities, entertainment, industry, business, in short, in various segments of society, including education, directly affecting the school teaching and learning process. For this reason, Valente reinforces this thought when he tells us that:

> Over the last few decades, the development of technology has taken on an increasing pace, taking society in new directions. Technologies are fundamental to the survival of our society, and since the invention of writing and the printing press, nothing like them has had such a social impact and stimulated so many changes. This means that new technologies affect many areas of society, including the organisation of educational systems and the teaching and learning process itself (VALENTE, 1999, p. 19).

According to Valente (1996, p. 16), the changes that have taken place in education, although **slowly , are due to the fact that** "new tools are being added to aid teaching, making students seek to obtain information and thus work on their **critical, observant side", but for this, the author tells**

us that, "we must have qualified professionals to be able to teach students to obtain information without **the need for them to be forced to learn what the teachers want to teach".**

But in order for the changes in education to expand and become more noticeable, it is necessary for changes to take place in schools as well. However, these changes are not so simple and involve the entire school team. According to Valente (1993, p. 22),

> Implementing new teaching techniques together with technology in the school environment is a way of revolutionising learning, where the teacher will no longer be a deliverer of information but will help the student to convert great information into knowledge that can be applied to solve problems that interest them.

Thus, in this educational context, students will feel motivated to seek out new knowledge techniques that will broaden and improve their skills and help them solve new questions.

In this way, we recognise that there are many benefits brought by technological resources to the educational environment and, for this reason, technology should be applied as a fundamental tool in this new teaching modality demanded by society, with the aim of developing skills that will be indispensable for students. But for this to happen, teachers need to be familiar with the tools at their disposal if they want learning to really happen, because the use of technology at school goes beyond making these resources available; it involves combining method and methodology in the search for more interactive teaching.

2.1 Technology and its integration into education

What is needed to integrate technological tools into the teaching and learning of students at different levels of schooling, as well as in the initial and continuing training of maths teachers? What do teachers need to know in order to teach effectively in technological contexts? What theoretical tools have been

built to study this question? Why do students have so many difficulties in understanding mathematical concepts? And how can technologies, especially computers, help with learning?

Kawasaki (2008) raises other questions: How can I use the computer to teach? In the specific case of maths teaching: How can I teach fractions using the computer? Or how to teach factoring? What about notational products? There are countless questions, **but, starting from the expectation of 'answers',** the author states that we could reformulate them as follows: What software/programmes exist to teach a specific content? Or how to incorporate a particular piece of software into the teaching of a specific piece of content?

Bittar (2010) also discusses some issues that teachers need to consider in order to integrate technology into their teaching practice, and not just insert it, as is often the case. In this sense, he makes a distinction between the integration and insertion of technology into teaching practice.

Insertion means what has been done in most schools: the computer is put in the school, the teachers use it, but it doesn't lead to learning that is different from what was done before and, what's more, the computer remains a strange instrument or alien to pedagogical practice, being used in unusual situations, out of class, that won't be evaluated.

Bittar (2010) argues that computers should be integrated into the school's pedagogical activities, i.e. they should be used and evaluated like any other instrument, be it chalk, concrete or any other material. And this use should be part of routine classroom activities. Thus, integrating a technological resource into teaching practice means that it should be used at various times during the teaching process, whenever necessary and in a way that contributes to the student's learning process. Integrating a technological resource into the teacher's pedagogical practice means that it becomes part of the arsenal available to the teacher to achieve their objectives.

There is no single way of describing the knowledge required of teachers to efficiently integrate technologies into maths teaching and learning, nor the correct way to prepare teachers for this task, nor a single way of carrying out technological integration. They are numerous and, above all, varied. They may differ in their formats, the content they cover, the computer tools they use, their objectives and the role they give to technology in the learning process. But all of them must implicitly have the expectation that they can overcome traditional teaching that does not achieve good results.

The appropriate use of ICT in the educational environment depends on the proposal to be used and the attitude of the professionals involved. Especially in maths teaching, there are many resources available, but there is still a lack of knowledge on the part of teachers in this area to make better use of these resources.

As we know, the arrival of ICT in schools has given teachers new roles. Learning how to work with the machine is no longer enough; you need to be aware of what's new and explore ways of making use of the potential of technology (COSTA and LINS, 2010). The authors also add that technology alone does not guarantee the quality of teaching and learning, but it can be a valuable resource through critical reflection on its limits and possibilities in collaborative work contexts.

Coan et al. (2013) state that the lack of technological resources, such as computer labs that are in full working order, is still a major obstacle for teachers and students, because when there is space, machines are missing or don't work. This is one of the factors that discourages teachers from investing in integrating ICT into the teaching process and planning their pedagogical activities in a way that makes it possible to use technology.

In our experience as educators, we have observed that the use of technological resources in public schools is often limited to internet searches and other online activities. The use of spreadsheets and other software goes almost unnoticed.

Martini and Bueno (2014), in a study on the frequency and purpose of ICT use in maths degree courses, state that ICT is not used regularly to work on specific content and that it is used more for out-of-class activities.

Paranhos (2005) carried out a study with primary school maths teachers who did not have access to technology during their initial training. This researcher analysed how the process of integrating these technologies into their lessons took place and identified the teachers' difficulties in this process. The main difficulties encountered were: not knowing how to use computers, the school's lack of structure and difficulties in designing motivating activities.

A more recent study by Marin (2012) is similar to the study by Paranhos (2005), pointing out that research has shown that the use of ICT is quite restricted, even though it can facilitate the study of various contents. The technical capacity of machines makes it possible to plan teaching activities that were previously unthinkable using blackboards and chalk.

Two hypotheses can be raised in relation to teachers' lack of use of technology. Firstly, many teachers don't have a good grasp of computers and some of their many tools, which can be used in favour of education. Secondly, educators don't like or don't want to use them.

In view of the first hypothesis, the most plausible justification may be that, due to the long working hours they are given, teachers don't have the time to prepare their lessons, nor to design activities that make use of computers, and especially to train themselves to do so.

With regard to the second hypothesis, Kawasaki (2008) states that in an inhospitable environment such as that of our maths teachers, it is unfair to say that teachers resist the use of new technologies. Therefore, in order to assess whether or not teachers are willing to make changes to their practice, it is necessary to take the focus off the individual and look at the context in which the teaching activity takes place.

Carneiro and Passos (2009) investigated the experiences of maths teachers at the start of their careers when using ICT in their classes and considered that the participants in the research had a great deal of creativity and diversity in the ways they used the technologies. In addition, most of the experiences they reported indicated that these tools were used from the perspective of an element of change, i.e. in order to carry out activities that would not have been possible without the use of technologies.

According to Mercado (2002), it is not enough to make computers available to teachers, but it is necessary to prepare them, respect their time and, above all, make them understand why they are using a new working tool.

Kenski (2007) states that schools must adapt to advances in technology and guide everyone towards mastery and critical appropriation of these new means. Thus, teaching becomes the means by which technological advances can be inserted into students' everyday school lives. Carneiro (2002), on the other hand, warns us about the importance of students and educators establishing a pleasurable commitment to new technologies and realising the need to abandon resistance.

For teaching maths, there is a lot of software that allows you to explore maths concepts in a more dynamic and detailed way. However, it is difficult to keep up to date when we talk about computer tools, more specifically software for any purpose, due to the quantity and speed with which these technologies and each of their new versions are developed. For this reason, from the point of view of the software applications available, there is no way of exhausting the discussion on this subject. Despite the rapid emergence of new technologies, we believe that very little maths software actually reaches the classroom. However, we have observed that the use of these applications in maths teaching and learning situations is restricted to a few groups of software.

Mathematical software is software that simulates mathematical processes. In other words, they perform arithmetic calculations, process and manipulate algebraic expressions symbolically and numerically, represent mathematical equations and geometric objects graphically.

Winplot is one of the many pieces of software available for free use and with potential for teaching maths. This software performs various functions, including plotting function graphs in two and three dimensions. It also allows the algebraic expressions of functions to be entered in explicit, implicit, parametric or polar form, which enables many interactions, use in various mathematical contents and at the most diverse teaching levels. It is also possible to make animations by defining the parameters of the functions studied.

According to Sousa (2013), the use of this software is suggested for five reasons: 1) It is entirely free; 2) It is simple to use, there is help in all parts of the programme and it accepts mathematical functions in a natural way; 3) It is very small and portable compared to existing programmes; 4) It is always up to date; 5) It is also in Portuguese.

Melo and Antunes (2002) state that "schools should encourage and adhere to the free software philosophy and promote an exchange of programmes and experiences in search of better learning" (MELO; ANTUNES, 2002, p. 65).

Like any other teaching resource, we don't want the software to be used in every lesson, but at appropriate times, with the teacher having the freedom to alternate the different resources depending on the objectives to be addressed in each lesson.

We agree with Borba and Villarreal's (2005) view that, in this conception of knowledge, there is no dichotomy between humans and technology. Knowledge is produced through constant interaction between humans and non-human collectives, in which the media reorganise thought. Every day, the use of information and communication technologies, especially the Internet, expands this interactivity. Machines process information, correct and translate texts, alter and edit images, and we know less and less where to place the boundary between them and our brains. From this perspective, the computer becomes one of the integral objects of our cognitive network, with which we have a healthy relationship of mutualism: it is an integral part of the web that allows us to act in the world.

Kenski **(2007) observes that "the computer** - and all its interactive possibilities for communicating and exchanging information - expands the quality and quantity of **information** consumption and **production" (KENSKI, 2007, p. 35). There is a new logic that influences the** ways of communicating and interacting with information, the production of knowledge and the ways in which memories are used.

Blikstein (2015) states that he finds it "strange", for example, to use the term ICT so often, forgetting that technology is not just for communicating and informing, but mainly for doing. ICT is an instrument of concrete making, of construction.

Valente (1999) reflects on this and identifies two practices that maths teachers can carry out when using technological resources, especially computers, in the classroom.

The first relates to activities using the computer only to transmit information, and consists of the computerisation of traditional teaching methods. The second is when the computer is used to help students construct new knowledge, making it a technological resource to be used by students, which enables them to describe problem-solving, reflect on the results obtained and deepen their ideas through the search for new content and new strategies (VALENTE, 1999).

Valente (1999) states that the computer can be an important tool, capable of passing on information to the user or facilitating the process of building knowledge. He also argues that by analysing software, it is possible to understand that learning (memorising or constructing knowledge) should not be restricted to the software, but depends on the student's interaction with it.

Santana and Borges Neto (2001) emphasise that: technology is not the centre of the didactic work, but the aim is to develop the student's critical and reflective sense of scientific knowledge. In this case, the computer would be an agent that would allow students to develop a new outlook.

According to Vianna and Araújo (2004), those in the classroom today cannot turn a blind eye to the use of information technology, because technologies linked to the practical activities of human beings produce behavioural changes. Thus, new skills and competences are needed when using these new technologies, in order to be able to develop knowledge, which, without a doubt, by using them, can broaden the concept of class, space, time and communication.

However, the use of technology as a tool in the teaching and learning process has been a challenge for teachers, as they face many difficulties, including the school's lack of structure in relation to new technological resources and, above all, the teacher's lack of training to use this resource, which is one of the relevant points in this process.

According to Bittar (2011), technology should be used to give students access to properties or aspects of a concept, as well as to enable them to engage in mathematical activities that are different from those usually dealt with in a paper and pencil environment. For example, Cabri-Géomètre allows students to explore a geometric construction by making conjectures based on the manipulation of the constructed figure, an activity made possible precisely by the software's dynamism.

ICT should be integrated into the school as a model that supports the teacher's methodology. It is therefore very important to evaluate the introduction and use of these resources, because the aim is not just to create digital content or turn ICT **into "teaching machines", but to provide the** teaching and learning process. Understand that ICT implies new ways of generating, mastering and disseminating knowledge.

In short, there seems to be no doubt about the importance of integrating technological resources into the pedagogical environment of the school, but not only that, integration must be thought of in such a way as to produce more learning, because technological resources add so many possibilities at the present time.

However, we realise that, even today, many teachers haven't managed to overcome some methodological paradigms that ignore the presence of this technology in the school environment, which could influence students' social lives, both inside and outside the school.

The most effective way of integrating technology into education is to modify or reform the Political Pedagogical Project (PPP) of educational units, inserting computers and educational software into these environments, which should be used constantly and not just in isolated projects. In order to do this, it is necessary for both the teacher and the student to have full access to these technological resources and to know how to use them in such a way as to contribute to the dynamisation and efficiency of the teaching and learning process.

Along the same lines, Valente advises us that,

> If we want Technology in Education to go beyond the limits of fads, we need to invest in transforming the school so that it can embrace new initiatives, thus helping to ensure that these proposals reach the very centre of the educational process in a meaningful way: the students (VALENTE, 1999, p.120).

So in order to integrate technology into education, we need to know Who is going to use the software? What it will be used for? How will it be used? So that this tool doesn't become just another resource without objectives and, consequently, doesn't achieve its educational purpose, which is to provide and energise student learning. For this reason, we must recognise that technologies enrich education and, consequently, bring significant changes to educational paradigms, bringing new teaching methodologies. Thus, this new teaching methodology, which uses technology as an educational tool, generates a significant production of knowledge and makes students, through their participation, achieve success in the classroom.

In view of this, Valente **tells** us **that** "the development of educational software has gained a great deal of momentum in recent years, bringing **new options to the market", which is why** "it is essential that a piece of software is appreciated in a practical situation of use". However, the author makes it clear **that "it is** the educator's pedagogical practice with their students that should guide their choice" (VALENTE, 1999, p. 111).

As a result, we can say that the integration of new technologies into education brings social and cultural changes, making old pedagogical models outdated in the face of new means of storing and disseminating information. Two great examples of this are: the use of computers, which inexplicably attract the attention of young people, making them develop a greater and better capacity to store the information provided; and internet resources such as forums, chats, blogs, among others, which give students the chance to express themselves, making their ideas and research visible to everyone.

Thus, we realise that technology enables students to develop cognitive skills and comes to the rescue of the current state of pedagogical deficiency in our school institutions, promoting the qualification and improvement of teachers and students, thus revolutionising dynamics and motivation, both when teaching and attending classes.

1.1 Pedagogical practice and technological resources at school

Student dissatisfaction with the traditional teaching model is notorious in every institution in this country. Lectures using only the blackboard and chalk no longer attract students' attention, so why not do things differently? And since students like and use technology so much, why not take advantage of it and use it to your advantage?

Faced with these questions, we see the need for schools to modernise and offer their students diversity, arousing the same enthusiasm as the games and films they usually watch, because it is already known that students learn from what grabs their attention. So the question is: are schools and teachers prepared for such a change? Unfortunately not yet, but both schools and teachers are gradually mobilising efforts to better understand the meaning and consequences of using new technologies in the school environment, because, according to Borba,

> For many teachers, the computer is a myth, i.e. there is the idea that it is a very powerful instrument that requires highly qualified people to handle it, which causes fear, insecurity and shivers on first contact. There is fear of the unknown, fear of showing incompetence to colleagues, fear of damaging the machine and causing harm, fear of not being able to develop computer skills (BORBA, 2007, p. 29).

However, many studies and researches have reflected on this thinking on the part of the teacher and also on this new pedagogical practice that is being demanded of the school, a practice that is gradually drawing up a teacher/student/content relationship, expanded in the process of learning how to learn through the use of technologies. In order to cope with this process, the school has expanded its remit. Its discussions and questions have turned to continuing teacher training and the use of ICTs. But there is still a need for a revolution in teacher training, as technology is something that has yet to be demystified for most teachers (CANDAU, 1991, p. 13).

Nowadays, there are numerous programmes that help teachers carry out interactive activities, but many teachers do not use these resources properly in the classroom. According to Candau (1991, p. 14),

> Using the computer in the classroom is the least of the teacher's challenges: it's using the computer in a way that makes the lesson more engaging, interactive, creative and intelligent that seems really worrying. Simply transferring a task from the blackboard to the computer doesn't change a lesson. It's essential that the methodology used is thought out in conjunction with the technological resources that modernity offers. The film, the interactive whiteboard, the computer, etc., lose their validity if the main objective is not maintained: learning.

With this in mind, we realise that the role of the teacher, when using technological resources, is to facilitate student learning and transform the content covered into something curious and interesting. For example, graphs and tables can be created in Excel with the help of the maths teacher, geometry can be worked on using educational software, seminars can be presented using slideshows or recordings with digital cameras, in short, countless interactive exercises which, if they are well guided, make the teaching and learning process easier and more dynamic.

So, faced with so many possibilities, it is worth noting that some authors Magalhães and Amorim (2003, p. 45) defends the idea that,

> We need to face our fears and use technological resources as support for our lessons, because we have to be sure that teachers will never be replaced by technology, but those who don't know how to take advantage of it run the risk of being replaced by others who do. Furthermore, the use of the Internet in the classroom provides support for teaching that is more centred on the student and their initiatives. It helps us to create new perspectives during lessons, and proves to be a useful tool for research projects, reader development and access to information.

Another very important aspect that should be considered when using technological resources is that they need to be challenging and provide students with an involvement with the world around them, because according to Vygotsky in his Theory of the Zone of Proximal Development, students need to be challenged today so that tomorrow they can learn effectively. In this theory, Vygotsky argues that what a child can only do with someone's help at the moment (when they are in the zone of proximal development), they will certainly be able to do on their own a little later (reaching the level of real development) (VYGOTSKY, 1984, p. 112).

In this context, and still taking Vygotsky's theory into account, we recognise the teacher as a mediator, whose role is to help the student leave the zone of proximal development and reach their real level of development, using technological resources and all the benefits they offer. Thus, regardless of the technological resource that the teacher will use, we realise that the teacher is the main player in mediating learning and making it more attractive and interesting for the students, because technological resources, as well as arousing the student's curiosity about what is being taught, help prepare them for a world that they are expected to know.

With all this in mind, we recognise that the benefits brought by technological resources to education are numerous and significant, but if teachers want learning to really happen, they need to be aware of the tools at their disposal, because the use of technology in the teaching and learning process goes far beyond making resources available, it involves combining method and methodology in the search for more interactive and effective teaching.

CHAPTER 3

GETTING TO KNOW GEOGEBRA 3D

GeoGebra is free software that makes it possible to work with various areas of maths in a dynamic and interactive way. With its broad approach and accessible language, it is possible to use it at various levels of education since it combines geometry, algebra and calculus in a single system, as stated by the GeoGebra Institute in Rio de Janeiro, which is part of the IGI (INTERNATIONAL GEOGEBRA INSTITUTES), which provides an excellent description of this application:

> Created by Markus Hohenwarter, GeoGebra is free dynamic maths software designed for teaching and learning maths at various levels of education (from primary to university). GeoGebra brings together geometry, algebra, tables, graphs, probability, statistics and symbolic calculations in a single environment. GeoGebra thus has the didactic advantage of simultaneously presenting different representations of the same object that interact with each other. In addition to the didactic aspects, GeoGebra is an excellent tool for creating professional illustrations to be used in Microsoft Word, Open Office or LaTeX. Written in JAVA and available in Portuguese, GeoGebra is cross-platform and can therefore be installed on Windows, Linux or Mac OS computers (2014).

GeoGebra is an application that makes it possible to carry out various constructions using its tools. With it you can study points, lines, segments, vectors, conic sections, as well as working with equations, coordinates, derivatives, integrals and many other possibilities.

The software has a user-friendly interface. As it is an application designed to help teachers, it is very user-friendly. As mentioned in its description, it has two different ways of representing the same object that interact with each other: the geometry window and the algebraic window. The geometry window is the part for constructed objects. Here you can make various changes, such as colouring objects, changing the thickness of lines, calculating areas and volumes, rotating and translating objects, displaying calculations, etc. The algebra window displays the algebraic representation of every constructed object.

GeoGebra also offers an exclusive text input area, which is used to write coordinates, equations, commands and functions in such a way that, by pressing the **enter** key, they are displayed in the geometric and algebraic window.

A very interesting feature is GeoGebra Pre-Release, where you can access the programme online, so you can use the programme without having to install it on your computer, tablet or mobile phone. As it runs on multiple platforms, students can use it at school, at home, at the internet café or on their smartphones, i.e. anywhere that has access to a computer connected to the internet and has the Java programming language installed, otherwise they can install it from the GeoGebra page itself.

The version of the software used in this research has an additional window compared to previous versions, which is the three-dimensional (3D) visualisation window. This visualisation window enables a three-dimensional display with additional tools for this function, while maintaining the same manipulation resources that previous versions of the software already had. This makes it easy and intuitive for users to significantly develop their geometric thinking and assimilate spatial notions in a simplified way that will help them solve problems.

The initial screen of the version used can be seen in figure 1.

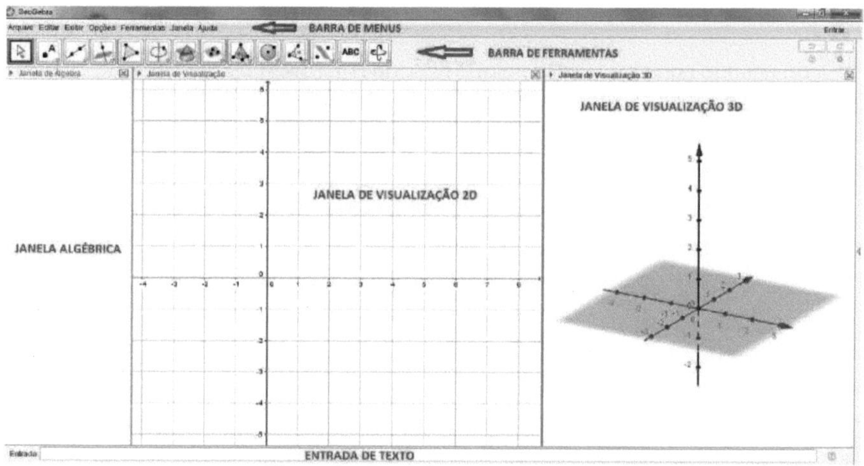

Figure 1 - GeoGebra 3D software interface

Source: GeoGebra 3D software *print screen*

As already explained here, GeoGebra 3D in its three-dimensional window works in a similar way to the 2D versions. This allows the user to carry out constructions using tools already pre-defined by the software itself. You can even build more elaborate and complex three-dimensional objects using relatively simple commands.

GeoGebra 3D has a vast number of icons on its toolbar (Figures 2 and 3). It is through these icons and/or by entering commands in the Text Input field that we can carry out various constructions of geometric figures, whether flat or spatial.

The icons shown in Figure 2 are triggered by clicking on the 2D Visualisation Window area, while the icons shown in Figure 3 are obtained by clicking on the 3D Visualisation Window.

Below we present in detail the functionality of some of the icons in GeoGebra 3D, highlighting those most commonly used when carrying out activities aimed at studying Spatial Geometry.

Figura 2 - GeoGebra 3D software's two-dimensional toolbar.

Source: GeoGebra 3D software *print screen*

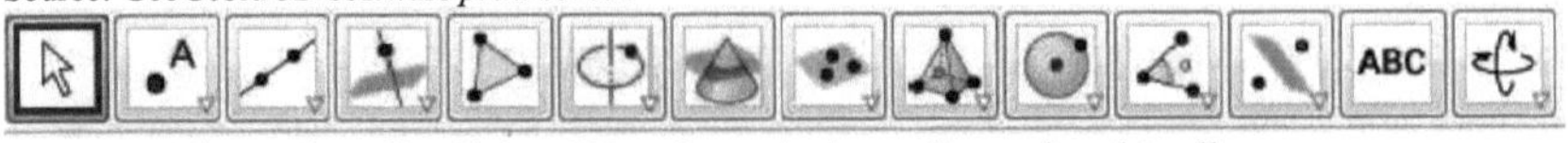

Figura 3 - GeoGebra 3D software's three-dimensional toolbar.

Source: GeoGebra 3D software *print screen*

1.  Icon used to select or move a constructed object.

2.  Icon used to construct a new point. This icon is followed by the following options:

Figura 4 - Options available for icon number 2.

Source: GeoGebra 3D software *print screen*

3.  Icon used to construct a line defined by two points. This icon is followed by the following options:

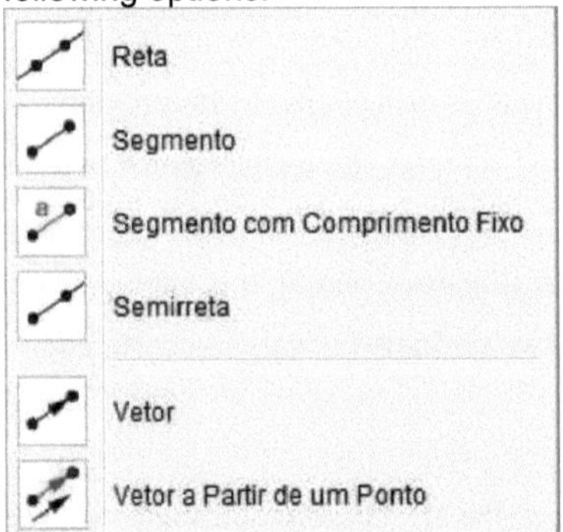

Figure 5 - Options available for icon number 3.

Source: GeoGebra 3D software *print screen*

4.  Icon used to construct a line perpendicular to a point and/or plane. This icon is followed by the following options:

Figure 6 - Options available for icon number 4.
Source: GeoGebra 3D software *print screen*

5. Icon used to construct polygons.

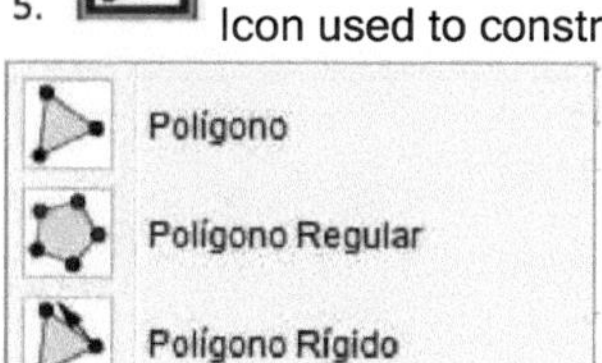

Figure 7 - Options available for icon number 5.

Source: GeoGebra 3D software *print screen*

6. 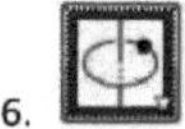

Icon used to construct a circle given axes and one of its points.

This icon is followed by the following options:

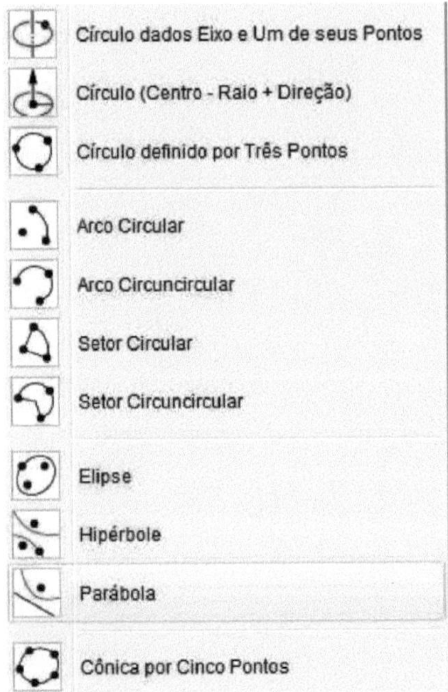

Fonte: *print screen do* software GeoGebra 3D
Figure 8 - Options available for icon number 6.

7. Icon used to construct intersection curves between two surfaces.

8. Icon used to construct a plane defined by three points. This icon is followed by the following options:

Figure 9 - Options available for icon number 8.
Source: GeoGebra 3D software *print screen*

9. Icon used to construct the main spatial figures such as: prisms, pyramids, cubes, tetrahedrons, cylinders, cones and more. This icon is followed by the following options:

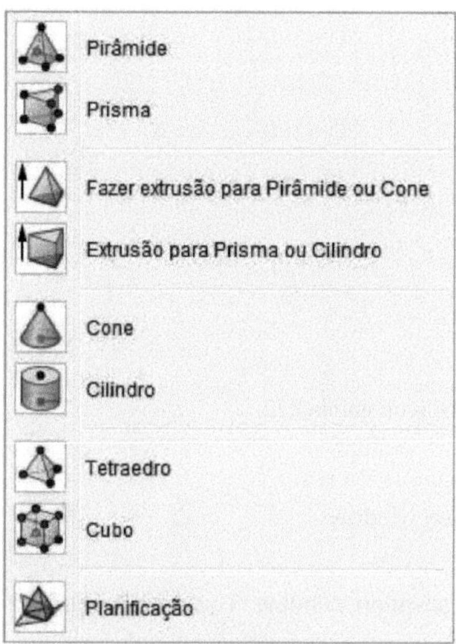

Figure 10 - Options available for icon number 9.
Source: GeoGebra 3D software *print screen*

10. Icon used to construct spheres given the centre and one of its points. This icon is followed by the following options:

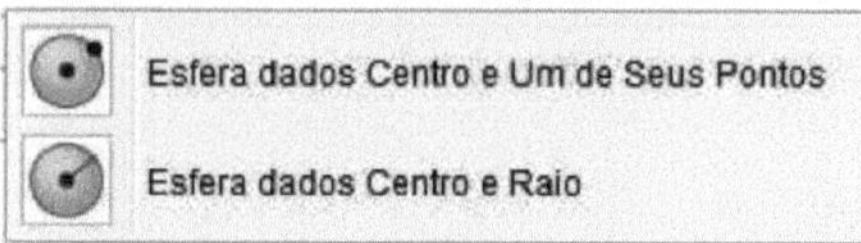

Figure 11 - Options available for icon number 10.
Source: GeoGebra 3D software *print screen*

11. Icon used to construct angles. This icon is followed by the following options:

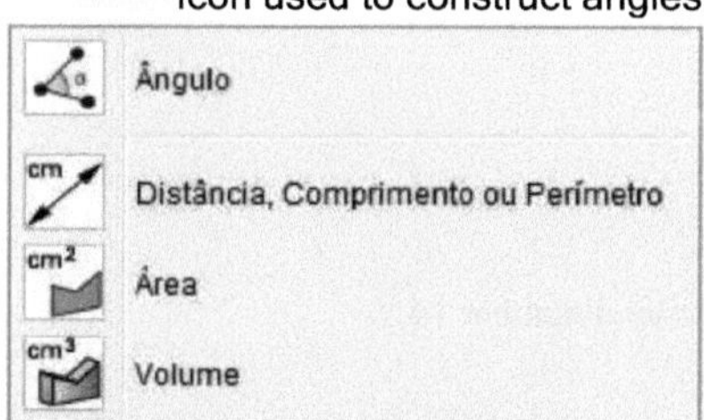

Figure 12 - Options available for icon number 11.
Source: GeoGebra 3D software *print screen*

12. Icon used for the reflection of two points by a plane. This icon is followed by the following options:

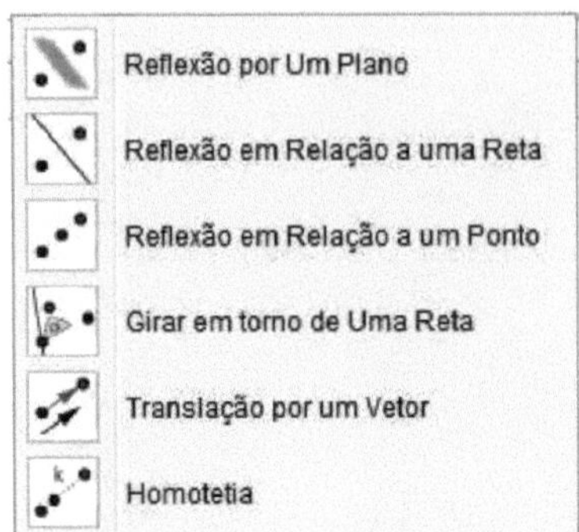

Figura 13 - Options available for icon number 12.

Source: GeoGebra 3D software *print screen*

13. Icon used to insert text into the construction window.

14. Icon used to drag and rotate the 3D Visualisation Window. This icon is followed by the following options:

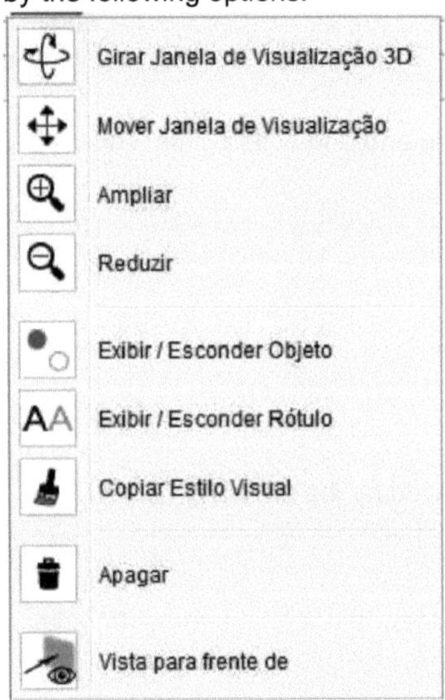

Figura 14 - Options available for icon number 14.

Source: GeoGebra 3D software *print screen*

This icon (slider) is only found in the 2D visualisation window and is used to define a variable. By defining this variable, we determine its range.

This diversity of tools offered by GeoGebra 3D allows the student to go far beyond the position of mere spectator, as the software provides the conditions and promotes the student to what we call the builder of mathematical knowledge.

The mathematical approach and didactic aspects achieved by GeoGebra strengthen the construction of investigative scenarios in which students are able to experience situations in a dynamic and interactive process, thus inviting them to discover, ask questions and seek answers so that the path that leads to knowledge is not characterised by being arduous and uninteresting, but rather pleasurable and meaningful.

CHAPTER 4

ON THE STUDY'S METHODOLOGICAL PATH

In order to carry out a study, we need to define and take ownership of the object we want to study, define the research subjects, select the data collection instruments and the theoretical contributions that support the discussions. In this chapter, we present some relevant points that characterise the methodological path of our work.

4.1 The steps of the study

In order to carry out the activities, it was necessary to use the following materials/tools: sulphite paper, textbooks, computers, the free GeoGebra 3D software, data-show.

The development of the proposal consisted of a set of activities spread over five meetings, totalling 10 hours/class, and dealt with the construction and analysis of geometric solids in the GeoGebra 3D software. The didactic sequence adopted can be seen in Table 1.

Chart 1 - Planning for the application of the didactic sequence

	ACTIVITIES	OBJECTIVES
Meeting 1 (2 h/a)	- Presentation of the research proposal; - Searching for and installing GeoGebra 3D - Handling the software tools	Familiarise students with GeoGebra 3D software.
Meeting 2 (2 h/a)	- Building prisms - Cube construction - Building pyramids; - Planning polyhedra.	Explore the main concepts involving the polyhedra worked on.
Meeting 3 (2 h/a)	- Cylinder construction; - Building cones.	Solving exercises from the textbook using solid constructions.
Meeting 4 (2 h/a)	- Solids of Revolution	Building the main solids of revolution.
Meeting 5 (2 h/a)	- Verification activity learning using GeoGebra 3D and the application of the questionnaire with the research subjects.	Check learning acquired with the aid of GeoGebra 3D and obtain the data to analyse the research.

Source: Prepared by the author.

E Meeting 1

The aim of the first meeting was to present the proposal developed in this research. All the students in the second year of secondary school, totalling 32, were present. Initially, the researcher informed the students of the objectives of the work, presented them with the timetable for carrying it out and the proposed activities that would be carried out. The students fully accepted the proposal. After answering the students' questions about the research, we began the process of searching for, downloading and installing the GeoGebra 3D software on their computers. The rest of the first meeting was dedicated to manipulating and discussing the characteristics and functionalities of each tool in the application to be used to study Spatial Geometry.

And Meeting 2

From this meeting onwards, we began to develop activities aimed specifically at the study of Spatial Geometry. So that the activities could be carried out with good results, the class was divided into groups of five and four students per machine. In this meeting we worked on the geometric solids prisms and pyramids.

Using the GeoGebra 3D software, we first built the prisms with triangular, quadrangular and hexagonal bases, similar to what can be seen in figure 15. Using the Rotate 3D Visualisation Window tool, we were able to identify and explore the basic elements of any polyhedron in the prism: vertex, edge and face.

In order to arrive at a definition of what a prism is, based on the students' own research and with the teacher's guidance, they came to the conclusion that a prism is a geometric solid made up of two convex polygonal regions, situated in distinct and parallel planes (called the base) connected by edges forming other faces (called the side faces) which are the parallelograms.

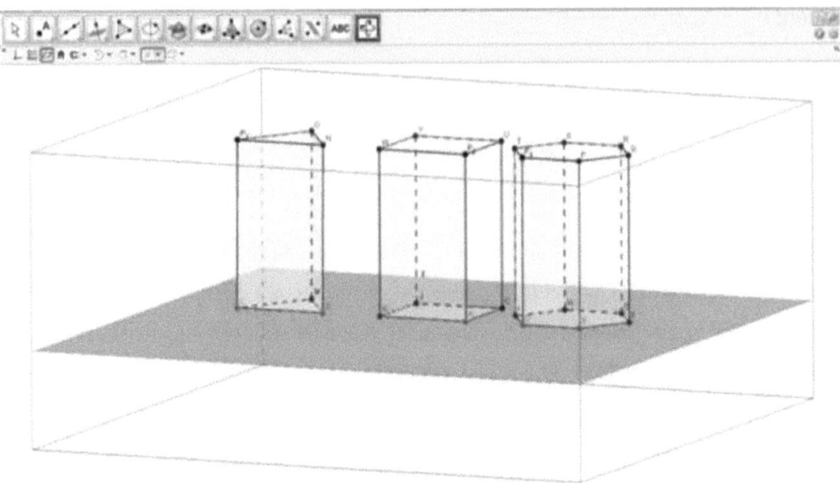

Figure 15: Triangular, quadrangular and hexagonal prisms built with GeoGebra 3D
Source: GeoGebra 3D software *print screen*

By manipulating the virtual objects, it is also possible for the teacher and students to take an approach involving Euler's relation, which through the formula V + F = A + 2 shows the relationship between the number of vertices V, edges A and faces F of a polyhedron.

By manipulating and visualising the virtual polyhedron, the students had the opportunity to verify the veracity of Euler's relation, where the application dynamically made it possible to obtain the data to verify this famous mathematical relation.

In order to provoke a discussion with the students about determining the volume of a prism, we built a square base where it could be seen on a horizontal plane. We assigned a height h along the vertical direction, so that as the height varied we built a different prism, as can be seen in figure 16.

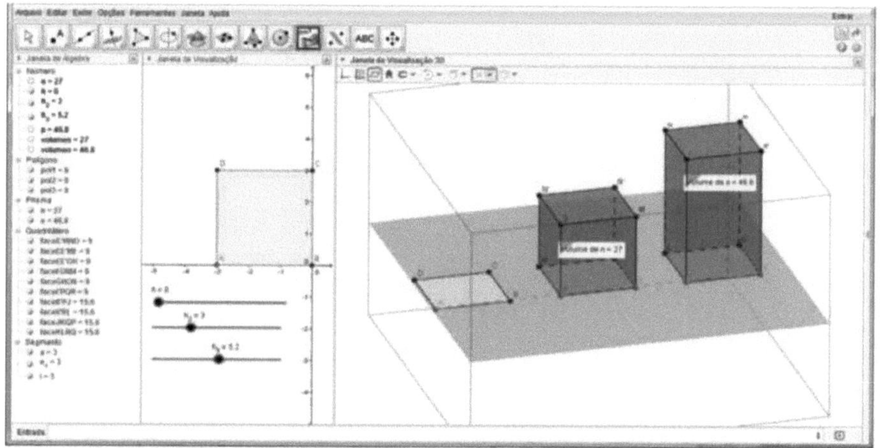

Figure 16: Quadrangular prism determining volume constructed with GeoGebra 3D
Source: GeoGebra 3D software *print screen*

It didn't take the students long to realise that the volume of a prism can be obtained by multiplying the area of the base of the solid by the size of its height h. By manipulating these virtual objects, it was possible to establish a generalisation with the students for calculating the volume of a prism, using the formula V = Ab x h, where Ab is the area of the base of the prism and h is the height.

When it comes to activities involving the planning of polyhedra, most of the students usually find it difficult to see the solid being planned, as they only have static figures presented in textbooks or even drawn by the teacher on the board. In an attempt to improve the students' understanding of the planarisation of polyhedra, the lesson explored the planarisation of the hexagonal prism, as shown in Figures 17, 18 and 19.

With the planned prism, exploring the concepts involved in calculating the area of this polyhedron became simpler and easier to understand. We can highlight that the dynamism offered by the software enabled the students to realise how geometric constructions have a direct relationship with the mathematical calculations carried out. This enabled the students to construct mathematical knowledge without being held hostage by ready-made formulas, thus characterising mathematics as a science under constant construction.

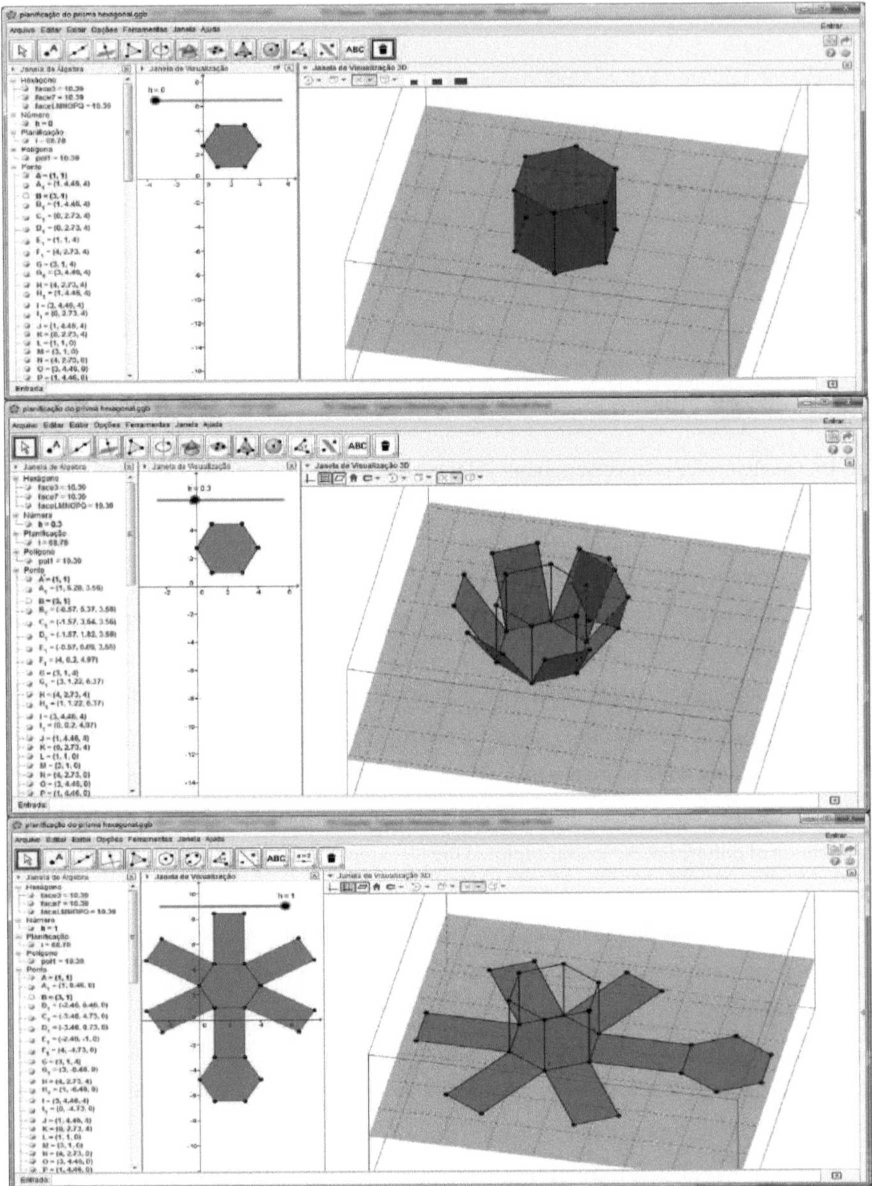

Figures 17, 18 and 19: Planning sequence of the hexagonal prism built with GeoGebra 3D

Source: GeoGebra 3D software *print screen*

During this meeting we also tried to explore the pyramid geometric solid. By this point, the students were showing greater mastery of the software, making it easier to construct the solids. Similar to what we did with the prism, we introduced the concepts that involve the pyramid solid, emphasising its basic elements (vertex, edge and face) and adding to our study the concepts of height and apothem, which are so important in Spatial Geometry. Using GeoGebra 3D and with the help of the teacher, the students built pyramids with triangular, quadrangular and hexagonal bases, as shown in figure 20.

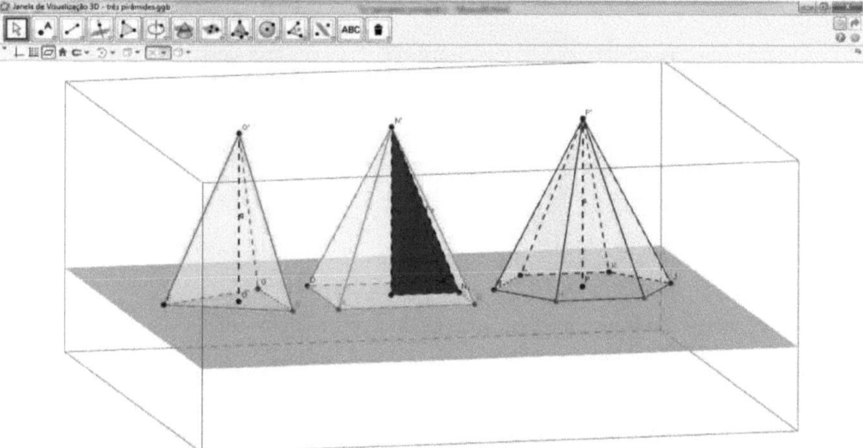

Figura 20: Triangular, quadrangular and hexagonal pyramids built with GeoGebra 3D

Source: GeoGebra 3D software *print screen*

The construction of these pyramids allowed the students, from a clear visualisation of the virtual objects, to directly relate the elements of the pyramid to the Pythagorean Theorem, using the relationship, $(\mathrm{ap}^{)(2)} = h^2 + (\mathrm{ab})^2$, where h is the height of the pyramid, (ab) is the apothem of the base and (ap) is the apothem of the pyramid.

We realise that using GeoGebra 3D software in Special Geometry classes offers teachers numerous advantages in various methodological aspects. We can mention, for example, the convenience of constructing geometric solids with maximum perfection

in a short space of time, enabling the teacher to interact more with the student. It is also important to emphasise the app's ability to approach content that is difficult to assimilate in a more playful and comprehensible way, thus helping students to acquire knowledge that was previously unattainable.

In the next activity, using the app, the students built three congruent prisms and inscribed pyramids in them, the bases of which were contained in different faces of the prisms and the vertices of which were common to the corresponding points of the polyhedron, as can be seen in figure 21.

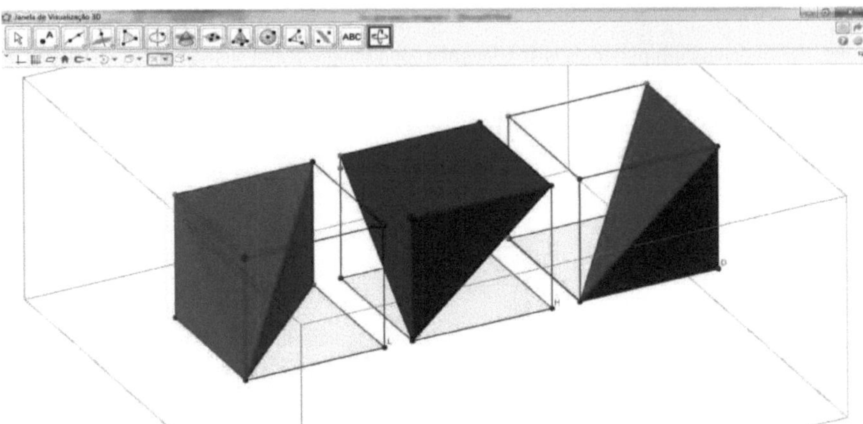

Figura 21: Congruent pyramids inscribed in corresponding prisms built with GeoGebra 3D

Source: GeoGebra 3D software *print screen*

By joining the three pyramids into a single prism, the students were able to see that they completely filled the volume of the polyhedron, as shown in figure 22.

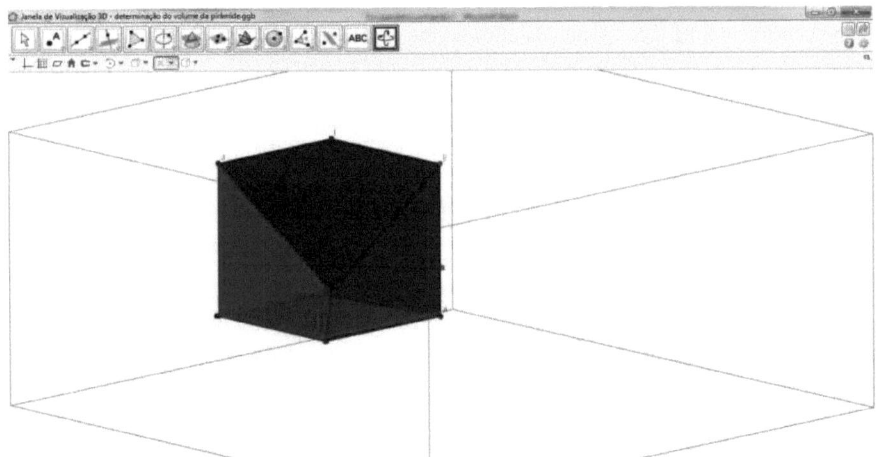

Figura 22: Solid formed from the junction of the three pyramids built with GeoGebra 3D

Source: GeoGebra 3D software *print screen*

In this way, it could be concluded that the volume of the pyramid corresponds to one third of the volume of the prism, thus giving greater meaning to this mathematical relationship.

In Geometry, visualisation is essential. It is often from visualisation that we can obtain the requirements that make it possible to understand algebraic formalism. Using the software, this activity made it possible to playfully deduce the formula for

1

The volume of the pyramid is given by: $V = \text{-}\ A_b \times h$, where, A_b represents the area of the base of the pyramid and h its height. This gave the students the opportunity to construct knowledge based on their own conjectures.

And Meeting 3

In this meeting, we used the students' textbook as a tool to search for questions about the geometric solids cylinder and cone. The aim of this activity was to use the GeoGebra 3D software to help visualise the geometric objects explored in the questions.

As the cylinder is one of the geometric solids most commonly found in everyday life, it allowed the teacher to use the construction carried out in class to explore problems that require students to exercise their intellectual capacity as well as finding strategies that contribute to logical-mathematical development.

Similarly, the geometric solid cone was explored by solving problems that contribute significantly to the construction of mathematical knowledge, thus enabling students to implement strategies that enable them to understand the content without being limited exclusively to the formalities of the subject.

In groups of 4 and 5 students, activities were carried out to solve problems involving the cone and cylinder solids. Using the software, the students were able to better explore the visualisation of the geometric objects worked on in the textbook problems. In this activity, we aimed to use the software as a facilitating resource for solving problems. Below we present some of the problems worked on in class and the solids built by the students based on their interpretation of each question. Figures 23, 24, 25 and 26.

01. A fuel tank is shaped like a straight circular cylinder with a diameter of 6 metres and a height of 15 metres. Determine the capacity, in litres, of this tank Figure ?: Cylinder from problem 1o. Use $\pi = 3.14$

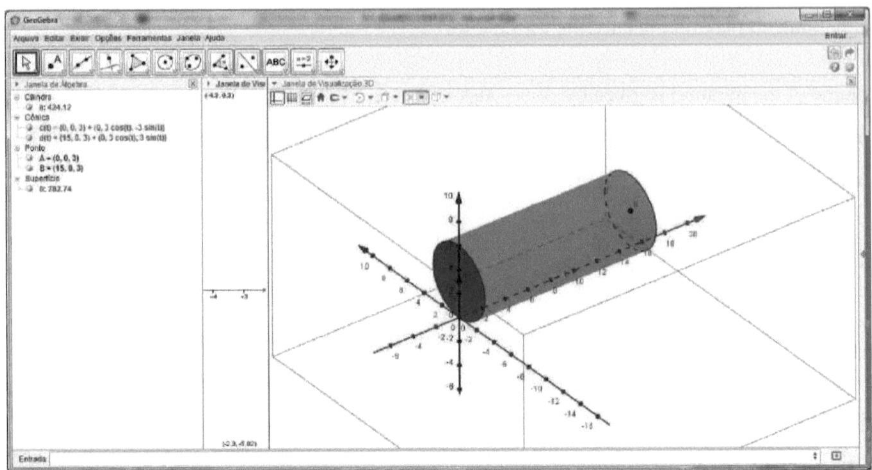
Figura 23: Cylinder for problem 1

Source: GeoGebra 3D software *print screen*

02. A cylindrical tank is 6 metres high and has a base radius of 2 metres. Determine the volume and capacity of this tank.

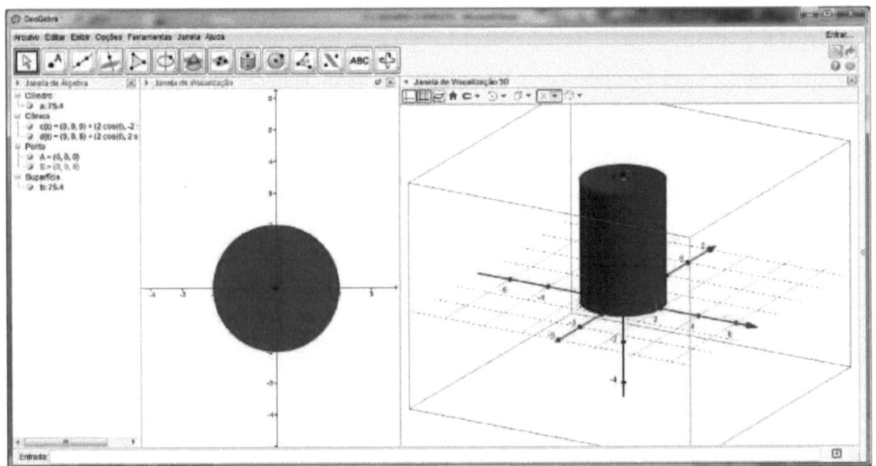
Figure 24: Cylinder for problem 2.
Source: GeoGebra 3D software *print screen*

03.(Vunesp - SP) An underground tank, shaped like a straight circular cylinder in a vertical position, is completely filled with 30 m^3 of water and 45 m^3 of oil. Considering that the height of the tank is 6 metres, calculate the height of the oil layer.

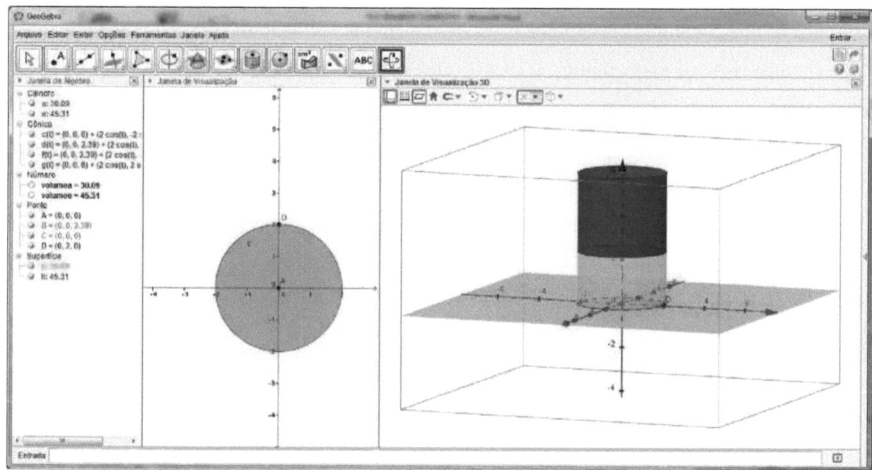

Figure 25: Cylinder for problem 3.

Source: GeoGebra 3D software *print screen*

04.Knowing that the luminaire should illuminate a circular area of 28.26 square metres, considering *n* = 3.14, the height h will be equal to:

a) 3m b) 4m c) 5m d) 9m e) 16m

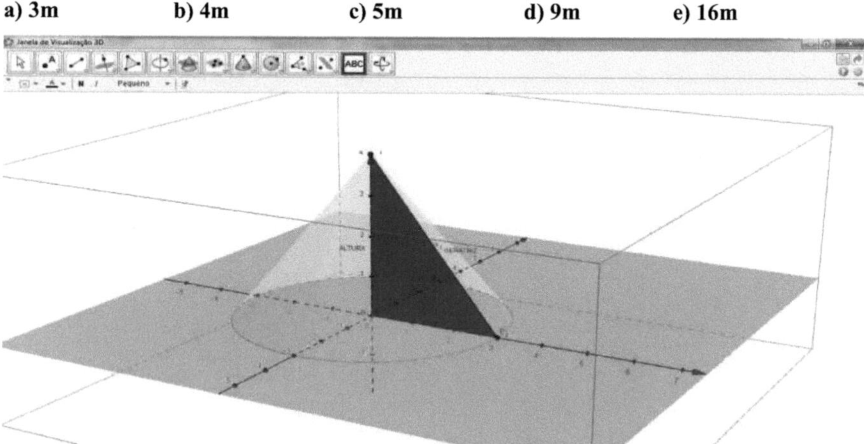

Figura 26: Cone for problem 4

Source: GeoGebra 3D software *print screen*

***And* Meeting 4**

In this meeting, using the GeoGebra 3D application, the students, with the help of the teacher, had the opportunity to build the main solids of revolution. We know that this is content that is hardly ever worked on with students in the classroom. However, in recent years we have seen the ENEM (national high school exam) present questions that demand success in solving them . The explanation for the absence of this content in classrooms probably lies in the fact that, in addition to the algebraic knowledge characteristic of mathematics, it requires the student to have great skill in visualising the geometric solids that originate from any flat figure.

So that the students could investigate the formation of some geometric solids, the teacher asked them to freely manipulate the application and check the results obtained.

The activity provoked immense excitement in the students. Curiosity and the satisfaction of discovery them to build several solids of revolution using GeoGebra 3D. Figures 27, 28 and 29 show some of the constructions made by the students.

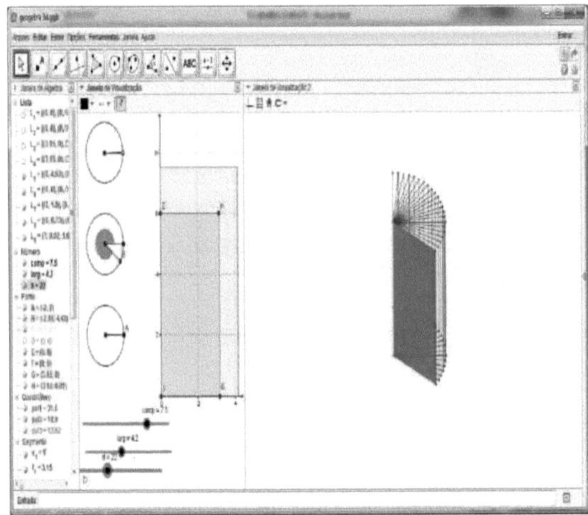

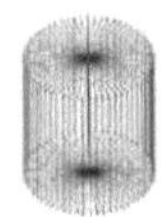

Figura 27: Revolution Cylinder

Source: GeoGebra 3D software *print screen*

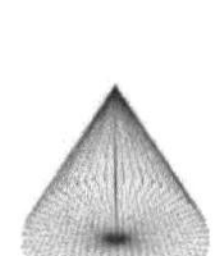

Figura 28: Cone of Revolution

Source: GeoGebra 3D software *print screen*

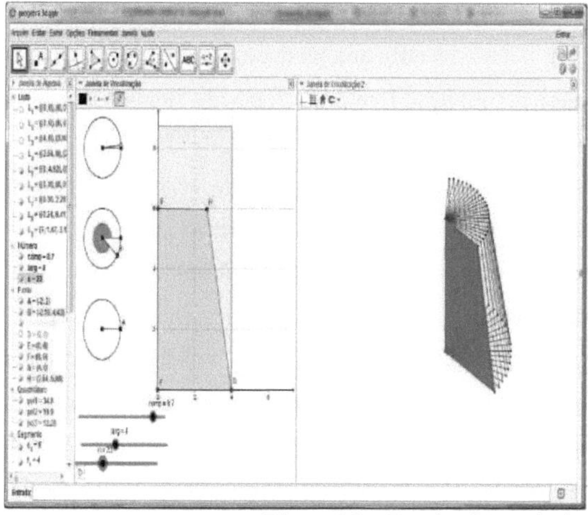

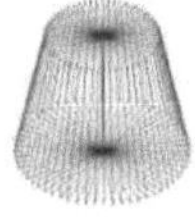

Figura 29: Trunk of Revolution Cone

Source: GeoGebra 3D software *print screen*

The constructions made with the GeoGebra 3D software gave us the opportunity to enter a world that is still little known in the students' mathematical universe. This was a simple activity that provided the students with entertainment, but with an immeasurable breadth of knowledge. The manipulation of the virtual objects allowed the students to empirically understand that the rotation of flat surfaces around an axis can generate countless solids, and that these are often present in our daily lives in various forms.

The most important thing about this type of activity is to realise that students acquire knowledge not through the act of memorising information, which often makes no sense to them, but through the pleasure of discovery.

And Meeting 5

In order to analyse the understanding of Spatial Geometry in its identification and classification points, calculations of areas and volumes of solids, a learning assessment activity was carried out in this fifth meeting.

For Dante (2013, p. 321),

> Assessment is a fundamental instrument for providing information on how the teaching-learning process as a whole is going -
>
> both for teachers and school staff to know and analyse the results of their work, and for students to check their performance.

In this sense, we sought to assess the student's ability to conceptualise, in such a way that they could indicate that they were able to explain, define, identify and produce examples and counter-examples that represent their concepts, because, according to Dante (2013, p.324) "Students can only give meaning to mathematics if they understand its concepts and meanings".

Based on this, we used the GeoGebra 3D software to show the knowledge produced and acquired. Once again, the class was divided into groups of four and five students. They had a week to carry out their activities, so that at the fifth meeting they could show all the knowledge they had acquired during this research using the Data Show.

As agreed, by the fifth meeting all the groups had their activities ready. With the collaboration of the other teachers who made their classes available to carry out the assessment, all the groups had enough time to present their conceptions and knowledge of the solids covered in the study of Spatial Geometry.

After carrying out the activity, we could see that a considerable number of the students were able to solidify their knowledge of the solids covered in this work. Others, on the other hand, were still cautious when expressing their ideas. However, they showed signs of evolution in terms of mathematical knowledge.

In general, we can say from the evaluation that the constructions in GeoGebra 3D helped significantly in the development of skills for visualising three-dimensional objects, as well as in understanding algebraic properties and relationships involving mathematical objects.

As happens in any educational environment, as in other social sectors regardless of the circumstances, the introduction of new methodological models is often accompanied by certain difficulties that need to be overcome. This study was no different: we faced difficulties such as an outdated computer lab, insufficient machines, students unsure of how to use the computer, a large class and other factors that influenced the execution of the didactic sequence. However, given the possibilities available, it was possible to overcome these difficulties and obtain positive results in the development of this research. As a result, we realised that implementing new teaching practices is no simple task. To

It is therefore necessary to have teachers who are prepared to break the barrier of traditionalism and include and integrate innovative resources that enable more effective and meaningful learning.

CHAPTER 5

GEOGEBRA 3D IN THE CLASSROOM

This chapter presents the results obtained from the questionnaires administered to the teacher of the subject and to the 32 students in the second year of secondary school at a public school.

The purpose of the questionnaire with the students was to analyse the contributions of the GeoGebra 3D software to the teaching of Spatial Geometry, and to check the opportunities offered for learning.

Faced with the question, **"Of the following factors, which ones most influence and hinder your learning of maths content?"**

Note: Students could tick more than one option.

( ) I have no difficulties

( ) I don't like maths

( ) Lack of teacher training

( ) Lack of dynamic resources in the classroom

( ) Too few lessons for too much content

() I don't like my teacher

( ) Other factors. Which factors?"

The results obtained for the students' answers were as follows:

N No difficulties: 6 students;

N I don't like maths: 19 students;

F Lack of teacher training: 1 student;

F Lack of dynamic resources in lessons: 7 students;

P Too few lessons for too much content: 8 students

N I don't like the teacher: 2 students

O Other factors. Which factors? 5 students.

Answers given by the students regarding other factors that influence their mathematical learning.

IR: "The lessons are tiring, I can't concentrate" (Student A)

R2: "I find maths a very **boring** subject" (Student B)

R3: "When I'm not at school I'm working. There's not much time left to study at home" (Student C).

R4: "The subject is very difficult" (Student D)

R5: "I've never learnt this subject and I don't think I ever will" (Student E)

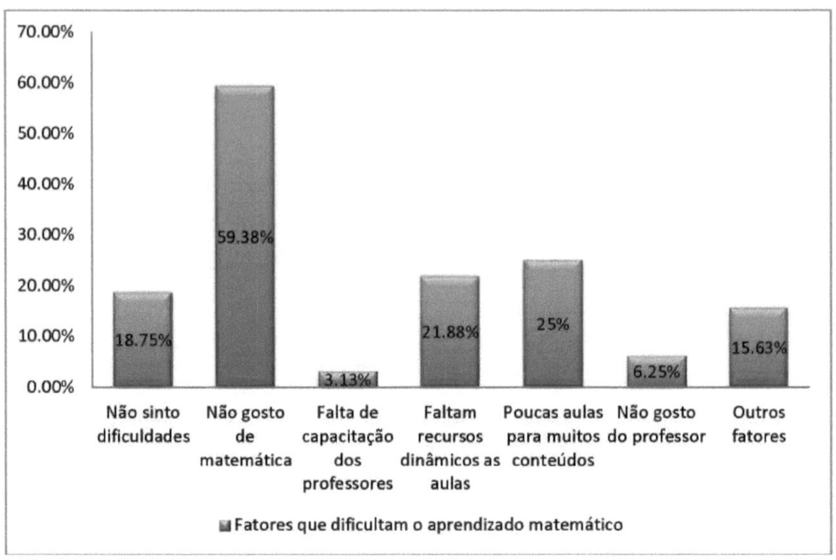

Graph 1 - Factors that hinder maths learning
Source: Prepared by the author

From the answers given, we can see that the main reason for students' difficulties in maths is that they don't like the subject due to a lack of understanding of the content being taught. This is clear from the words of Silveira (2002) when he tells us that students' dissatisfaction is expressed by "maths is boring", "I don't like maths", "maths is difficult". This data shows a worrying factor that has been perpetuated for decades in the school environment: rejection of the subject of maths. As emphasised by Silveira (2002), rejection of this subject has a direct influence on the teaching-learning process and on previously formed opinions. One point that can be considered positive in the results is that the students themselves, perhaps even unconsciously, indicate the path to be followed so that we can reverse this situation. Indirectly, they point out that if there is no change in methodologies, there will be no change in results, something that seems obvious but is disregarded by many who remain silent about the situation. The same data shows us that it is no longer possible to maintain yesterday's classroom patterns with today's students. Society has evolved, and school education as a social product needs to keep pace with this evolution. In the case of maths, it is necessary to leave the static environment and enter the more dynamic world. To do this, it's up to each teacher to review their methodological behaviour. And since we live in contemporary society, implementing technological resources in maths teaching seems to be a good starting point.

As such, there is a very strong tendency in the students' answers to say that they don't like maths. In many studies, this position predominates. However, a study by Baraldi (1999) shows that depending on the methodology used by the teacher, this understanding of maths changes. According to this author, problem-solving and dialogued lectures provide a dynamic and meaningful glimpse of mathematics and are capable of highlighting, even in an incipient way, its constructive aspect.

To the question **"Do you think the use of technological resources in class can help *you* improve your performance in the subjects you study?"** the results were as follows:

S Yes; 30 students

N No; 00 student;

T Maybe; 02 students

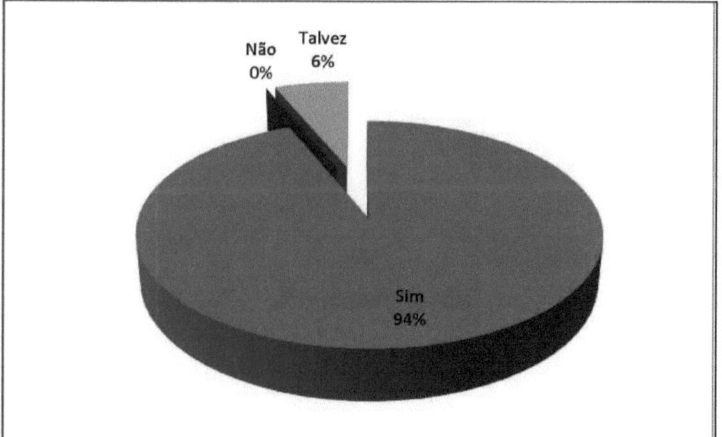

Graph 2 - Contribution of technological resources to subject performance

Source: Prepared by the author

From the answers given, we can see that the students consider technological resources to be a methodology that helps their performance and their learning. In view of this, we can see" the students' dissatisfaction with the traditional teaching model" and the **fact that** "lectures using only the blackboard and chalk no longer attract the students' attention". (VALENTE, 1999). It is clear from these responses that the use of technology is accepted, as it enables students to develop cognitive skills and comes to the rescue of the current state of pedagogical deficiency in our school institutions, promoting the qualification and improvement of teachers and students, thus revolutionising the dynamics and motivation of both teaching and attending classes.

With the question **"Does the maths teacher usually use the computer lab in their lessons?",** we obtained the following result:

S Always: 00 student;

À Sometimes: 32 students;

D Hardly: 00 student;

N Never: 00 student.

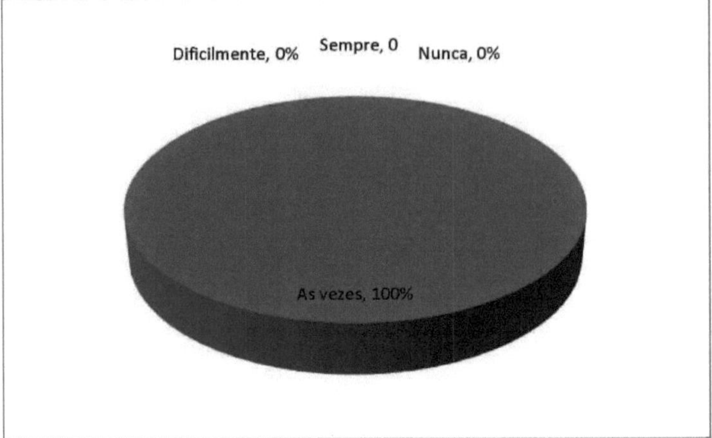

Graph 3 - Frequency of use of the computer lab

Source: Prepared by the author

Thus, through the students' responses, we detected that there is no constant use of technological resources during maths lessons, but it must be recognised that, even sometimes, these resources are being worked on and can be expanded more and more. To this end, Magalhães and Amorim (2003) suggest that "we need to face our fears and use technological resources as support for our lessons". They also emphasise that "teachers will never be replaced by technology, but those who don't know how to take advantage of it run the risk of being replaced by others who do". We must also increasingly recognise that "the use of the Internet in the classroom provides support for teaching that is more centred on the student and their initiatives", **and that, as well as "creating new** perspectives during lessons, it is proving to be a useful tool in the area of research for projects, reader development and access to information" (MAGALHÃES and AMORIM, 2003, p. 45).

After getting to know the GeoGebra 3D software, the students were asked **"How do you classify the purpose of using the GeoGebra 3D software?", and** we obtained the following result:

M Very useful: 21 students;

Useful: 09 students;

P Not very useful: 02 students.

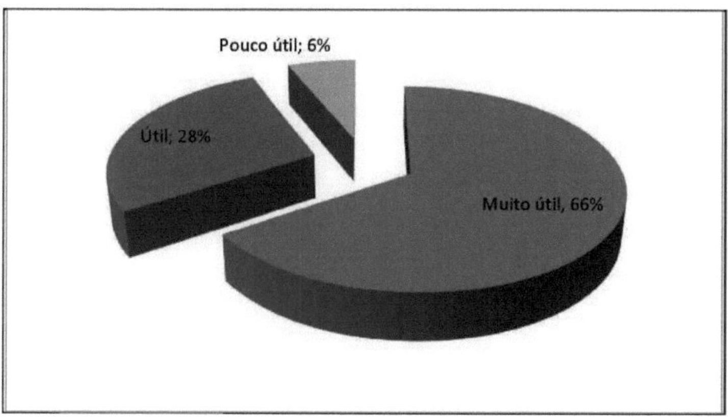

Graph 4 - Purpose of using GeoGebra 3D software

Source: Prepared by the author

The results achieved with the answers given by the students show their acceptance of the use of the software in question and, if they found the software useful, it is consequently because it helped them to understand the mathematical content studied. In view of this, we can say that the software fulfilled its objective, which is to work on maths in a dynamic and interactive way.

We also asked the students **"How do you rate the visualisation of solids using GeoGebra 3D?"** and the answers we got were:

M Improved a lot: 27 students;

M Improved: 03 students;

N No difference: 02 students.

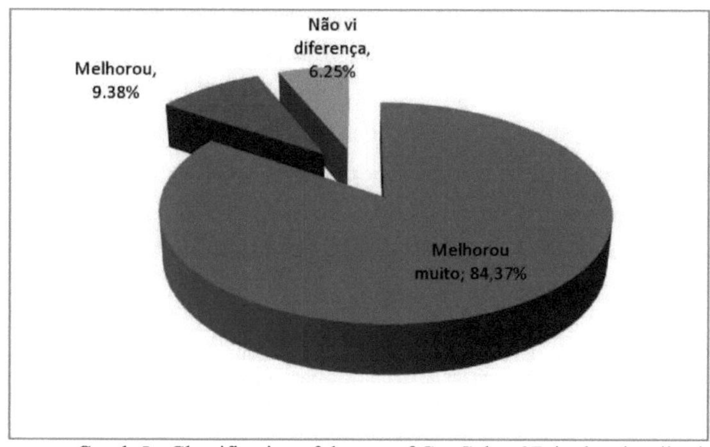

Graph 5 - Classification of the use of GeoGebra 3D in the visualisation of solids

Source: Prepared by the author

In relation to the **question "What has changed in terms of learning the content worked on using the software?"**, the students replied that:

F Made it much easier: 21 students;

F Partly facilitated: 09 students;

N Nothing has changed: 02 students.

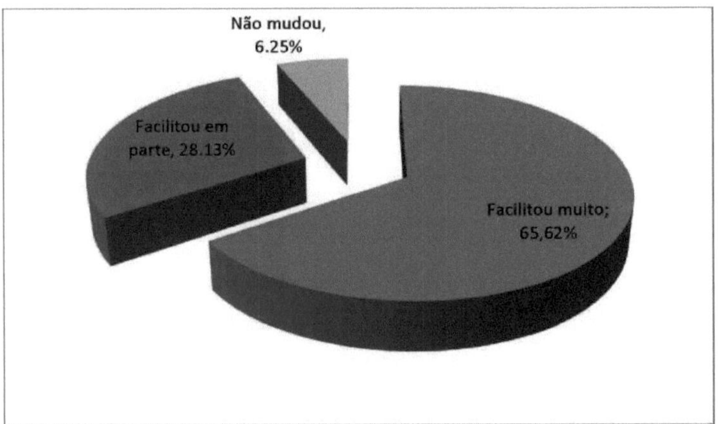

Graph 6 - Learning content using the software

Source: Prepared by the author

The results achieved with the answers given by the students in questions 4, 5 and 6 show their acceptance of the use of the software in question and, if they found the software useful, it is consequently because it helped them understand the mathematical content studied.

Based on the answers given, we can say that the software has fulfilled its objective of working with maths in a dynamic and interactive way. Through the software, students were able to experience situations in a dynamic and interactive process, thus being able to discover, formulate questions and seek answers in a pleasurable and meaningful way.

Thus, through the data collected in this research, it can be seen that the use of computer programmes in maths classes provides benefits that would be difficult to achieve without their help. In Geometry, especially Spatial Geometry, the visualisation factor is essential for students and teachers to be able to make their own conjectures. The construction of geometric thinking begins with the visualisation of space and its shapes. According to the PCN,

> Geometric thinking develops initially through visualisation: children know space as something that exists around them. Geometric figures are recognised by their shapes, by their physical appearance, in their entirety, and not by their parts or properties (BRASIL, 1998, p. 127).

A similar thought can also be observed in question 7 of the questionnaire, which asks students to **"Give your opinion on the use of GeoGebra 3D software in the teaching of Spatial Geometry".** We observed that the students generally responded that the software helped them understand the content and that they were able to assimilate the content better and not get tired while studying. One response that makes this clear is that of Student E, who expressed his opinion on the use of GeoGebra 3D software in the teaching of Spatial Geometry as follows:

> I really enjoyed studying this subject in this way, using the computer programme to turn the figures in any direction we wanted. The subject studied in this way was easier to understand. It didn't even seem like I was studying maths (Student E).

After analysing the questionnaire administered to the students, we turned our attention to analysing the questionnaire administered to the maths teacher who, in order not to be identified, we will refer to by his profession, i.e. just as a teacher. This teacher has a degree in maths, has worked in this profession for ten years and currently teaches secondary school classes.

When we asked the teacher **"What preparation do you think teachers need in order to carry out their educational activities using the new technologies available at school?"**, the teacher gave us his opinion, revealing that,

> The use of new technologies in educational practice cannot be done irresponsibly and without objectivity, because any teacher must, first and foremost, have a good understanding of the pedagogical nature of the technologies in order to be able to use them. For this reason, I recognise that teachers must undergo training that addresses the importance and limits of these resources in the pedagogical universe, as well as their physical structure and handling, from the most basic notions to the most advanced functions.

By thinking in this way, the teacher is taking a position similar to that of Valente (1993) when he reveals that technology should be applied as a fundamental tool in this new type of teaching demanded by society, with the aim of developing skills that will be indispensable for students. However, he also points out that for this to happen, teachers need to be familiar with the tools at their disposal if they want learning to actually take place, because the use of technology at school goes beyond making these resources available; it involves combining method and methodology in the search for more interactive teaching.

When asked **"What knowledge do you believe is necessary for teachers to use new technologies at school in order to achieve meaningful learning?"**, he revealed that,

> Teachers need to be aware of their students' realities and understand and master the pedagogical potential of these tools, because you can't just use them for the sake of using them. You need meticulous planning linked to research in order to know which

> points of the subject can be worked on, as well as their functions and resource limitations.

With this response, the teacher offers more elements to what Candau (1991, p. 14) says, when he reveals that using the computer in the classroom is the least of the teacher's challenges; he must use the computer in such a way as to make the lesson more engaging, interactive, creative and intelligent. The author also argues that it is essential that the methodology used is thought out in conjunction with the technological resources that modernity offers.

The film, the interactive whiteboard, the computer, etc., lose their validity if the main objective is not maintained: learning (CANDAU, 1991, p. 14).

To the question **"Do you believe that the use of technology can favour meaningful learning? Why?"**, the teacher told us yes, because, according to him,

> We live in a time of great technological advances and it is well known that new technologies are being used every day to improve the performance and understanding of activities carried out in various areas of human knowledge, such as medicine, architecture, engineering, etc. Given all this, and thinking about the school world, the challenges faced by teachers today in managing to hold students' attention for what they propose to work on in the classroom, we realise that they are not doing well in this fight. Faced with the fact that the reality experienced by students is already part of a world surrounded by technological innovations that manage to kidnap their interest, pleasure and curiosity, teachers will have to urgently review their teaching practices in order to put them in a position to compete with these novelties used by students and which even get in the way of a good lesson. They can no longer assume a position of resistance to innovations in the teaching-learning process and, like other human activities, must bring the virtual and technological world into the educational environment with the aim of helping students' cognitive performance. In other words, the school institution necessarily needs to find ways and means of using Information **and Communication** Technologies **(ICTs) in** educational **processes**, given that this is what currently manages to attract students' interest.

By positioning himself in this way, the teacher's thinking is similar to Valente's (1993, p. 22), when he tells us that new teaching techniques need to be implemented in the school environment along with technology, because this is a way of revolutionising learning, and with it the teacher will stop being a deliverer of information and will start helping the student to convert great information into knowledge to be applied to solving problems of interest to them.

In light of the above, we realise that the use of GeoGebra 3D software contributes to making the teaching and learning process more dynamic, stimulating students' performance and interest in mathematics and making a positive contribution to the development of learning.

CHAPTER 6

CONCLUSIONS

The work carried out here discussed issues related to the use of technology in the teaching of Spatial Geometry. We sought to analyse the contribution that the GeoGebra 3D software offers to teachers and students so that both can succeed in the process of teaching and learning mathematics.

We can see that GeoGebra 3D offers numerous advantages for student development, giving rise to greater interest in learning. With it, the teacher has the opportunity to explore various areas of maths, from the simplest to the most complex, enabling the student to better understand the construction of new mathematical concepts.

This experience has allowed us to reinforce the thesis that students like different activities and that these help to facilitate learning. The activities developed with the GeoGebra 3D software show that it is possible to teach geometry in a dynamic way, making the lesson exciting and attractive, in which the student participates, interacts with his classmates, and through his constructions formulates his own knowledge.

The use of dynamic geometry software enables students to progress in their mathematical thinking through the construction of geometric solids. It enables them to formulate conjectures through direct manipulation with mathematical objects.

To this end, it is necessary to carefully analyse the software to be used in the development of educational activities. In such a way that the educator is able to achieve their objectives in relation to the curricular requirements of each subject. Educators need to be aware of the resources offered by the application. As well as identifying potential in the software that can contribute to better student performance in everyday school life.

It's worth pointing out that learning doesn't just happen through the implementation and use of technology in the classroom. However, we believe that didactic work using various technological resources enables better learning of mathematical content.

Based on the actions and activities developed in this study, it is possible to see that the use of new technologies, especially the GeoGebra 3D software, can "awaken" students' curiosity and interest in understanding mathematical content that was previously considered the privilege of a few, especially Spatial Geometry, knowledge that is so important in our daily lives.

Therefore, based on the data found in this study, it is possible to conclude that the GeoGebra 3D software can make a significant contribution to improving the teaching of Spatial Geometry, especially in terms of visualising the properties of solids. In addition, it is worth considering that the use of technology in classroom activities promotes an education focused on the quality and significance of the educational process as a whole, developing aspects and characteristics that will be fundamental to the student's development and their inclusion in the technological universe.

REFERENCES

ALRO, H.; SKVSMOSE, O. **Dialogue and learning in mathematics education**. Belo Horizonte: Autêntica, 2006.

BRAZIL. Ministry of Education and Sports. **National Curriculum Parameters: Maths.** Brasília: MEC/SEF, 1998.

BARALDI, I. V. Maths at school: what science is this? Bauru: EDUSC, 1999.

BITTAR, M. A Escolha do Software Educacional e a Proposta Didática do Professor: estudo de alguns exemplos em matemática. In: BELINE, W.; COSTA, N. M. L. (Org.). **Educação Matemática, Tecnologia e Formação de Professores**: algumas reflexões. Campo Mourão, PR: Editora de Fecilcam, 2010, v. único, p. 215-243.

__________ The instrumental approach to the study of the integration of technology in the pedagogical practice of maths teachers. **Educar em Revista**, Curitiba (PR), n. Especial 1/2011, p. 157-171, 2011. UFPR Publishing House.

BORBA, M.C. **Informática e Educação Matemática.** Belo Horizonte: Autêntica, 2007.

BORBA, M. C.; VILLARREAL, M. Humans-with-Media and the Reorganisation of Mathematical Thinking: information and communication technologies, modeling, experimentation and visualization Mathematical. New York, USA: Springer, 2005.

CARNEIRO, R. F.; PASSOS, C. L. B. Experiences of early-career maths teachers in the use of information and communication technologies. **Zetetiké** - Cempem - FE -Unicamp - v. 17, n° 32, jul/dez, 2009.

CARNEIRO, R. Informática na educação- representações sociais no cotidiano. São Paulo: Cortez, 2002.

COAN, L. G. W. et al. ICT in maths teaching: teacher training under debate. REVEMAT, ISSN 1981-1322. Florianópolis (SC), v. 08, n° 2, p. 222-244, 2013.

COSTA, M. L. C.; LINS, A. F. Collaborative work and the use of information and communication technologies in maths teacher training. **Educ. Matem. Pesq**., São Paulo, v.12, n.3, pp. 452-470, 2010.

CANDAU, V. M. F. **Informática e educação:** um desafio. Educational Technology. Rio de Janeiro, v. 20, p. 14-23, jan./abr. 1991.

DANTE, L. R. **Matemática: contexto e aplicações.** 2ª Ed. São Paulo: Ática, 2013.

FERNANDES, E. (org.) **Technology needs to be in the classroom.** PUC-SP researcher talks about technology in the classroom. Revista Nova Escola [on-line], n° 233, 2010. Text available at: http://gestaoescolar.abril.com.br/aprendizagem/entrevista-pesquisadora-puc- sp-tecnologia-sala-aula-568012.shtml/. Accessed on 04 November 2014.

SOUZA JUNIOR, M. L. S.; BARBOZA, P. L.**Pathways in the pedagogical practice of**

mathematics.REVEMAT, ISSN 1981-1322. Florianópolis (SC), v. 08, n. 1, p. 199-215, 2013.

GOMES, C. L. (org.). **Critical Dictionary of Leisure**. Belo Horizonte: Autêntica, 2008.

GeoGebra Institute in Rio de Janeiro. Available at: <http://www.geogebra.im- uff.mat.br/>. Accessed on 14 November 2014.

KAWASAKI, T. F. **Technologies in the maths classroom**: resistance and changes in continuing teacher education. Thesis presented to the Postgraduate Programme in Education at the Faculty of Education of the Federal University of Minas Gerais, 12 December 2008.

KENSKI, V. M. **Educação e tecnologias**: o novo ritmo da informação. Campinas, SP: Papirus, 2007.

LORENZATO, S. Laboratório de ensino de matemática e materiais didáticos manipuláveis.In: LORENZATO, S. (Org.) **Laboratório de Ensino de Matemática na formação de professores**. Campinas (SP): Autores Associados, 2006. p. 3-38.

MAGALHÃES, V.; AMORIM, V. **Cem aulas sem tédio**. Porto Alegre: Instituto Padre Reus, 2003.

MARIN, G. Advantages and disadvantages pointed out by maths teachers in the use of technology in the use of information technology in higher education. **Electronic Journal of the Teacher Training Division** (http://www.seer.ufu.br/index.php/diversapratica) v. 1, n. 1 - 2nd Semester 2012.

MARTINI, C. M.; BUENO, J. L. P. The challenge of information and communication technologies in the initial training of maths teachers. **Educ. Matem. Pesq.**, São Paulo, v.16, n.2, pp. 385-406, 2014.

MOTTA, Carlos Eduardo Mathias. **New technologies in maths teaching.** Rio de Janeiro: UFF/ Fundação CECIERJ, 2004.

MOYSÉS, L. **Aplicações de Vygotsky à educação matemática**. 10ª Ed. Campinas: Papirus, 1997.

MADRUGA, J. A. G. **Learning by Discovery versus Learning by Reception:** The Theory of Significant Learning. In: COLL, C.; Palácios, J.; MARCHESI, A. (Orgs.). Psychological development and education: psychology of education. vol. 2. Porto Alegre: Artes Médicas, 1996.

MARTINI, C. M.; BUENO, J. L. P. The challenge of information and communication technologies in the initial training of maths teachers. **Educ. Matem. Pesq.**, São Paulo, v.16, n.2, pp. 385-406, 2014.

MELO, M. M. M.; ANTUNES, M. C. T. Free software in education. In: MERCADO, L. P. L. (Org.). New technologies in education: reflections on practice. Maceió: Edufal, 2002.

MERCADO, L. P L. (Org.). New technologies in education: reflections on practice. São Paulo: Edufal, 2002.

NETO, J. M. **As Pesquisas Sociais.** Ciência e Educação, Bauru: v. 9, n. 2, p. 147-157, 2003.

PARANHOS, L. R. L. From possibility to reality: an action research on the training of reflective and autonomous teachers in the use of information technology in education. 2005. Dissertation (Master's in Education) - Federal University of Mato Grosso do Sul, 2005.

SILVEIRA, Marisa Rosâni Abreu. **"Maths *U* difficult":** A preconstituted meaning evidenced in students' speech.2002. Available at: <http://www.anped.org.br/25/marisarosaniabreusilveirat19.rtf>. Accessed on 29 November 2016.

SOISTAK, M. M.; PINHEIRO, N. A. M.; PILATTI, L. A. Analysing the work carried out in the municipal public schools of Ponta Grossa in the teaching of mathematics through interdisciplinary projects. **VIDYA**, v. 31, n. 2, p.25-40, jul./dez., 2011 - Santa Maria, 2011.

SOUSA JÚNIOR, M. L.; BARBOZA, P. L. Pathways in maths teaching practice. **REVEMAT**. Florianópolis (SC), v. 08, n. 1, p. 199-215, 2013.

SOUSA, A. L. N. **Analysis of students' conceptions of maths teaching using Winplot at three levels:** primary II, secondary and higher education. 2013. 106fls. Dissertation (Master's in Maths) - State University of Southwest Bahia (UESB). Vitória da Conquista, 2013.

SANTANA, J. R.; BORGES NETO, H. Introduction of new technologies in maths teaching: continuing teacher training at the NTE in Quixadá in August 2000. Congress; **XV Meeting of Educational Research in the North and Northeast**; UFMA; São Luis/MA.2001.Available at. in: http://www.multimeios.ufc.br/arquivos/pc/congressos/congressos-introducao-denovas- tecnologias-no-ensino-de-matematica.pdf. Accessed on: 15, 11 April 2017.

VALENTE, J. A. **Teacher training:** different pedagogical approaches. In: VALENTE, J. A. (Org.). O computador na sociedade do conhecimento. Campinas: NIED, 1999. chap. 6, p. 131-156.

_____. **The Teacher in the Logo Environment:** training and performance. Campinas: UNICAMP/NIED, 1996.

_____. **Computers and Knowledge:** rethinking education. Campinas: Gráfica Central da UNICAMP, 1993.

VYGOTSKY, L.S. **Social Formation of the Mind**. São Paulo: Martins Fonte, 1984.

_______. Learning and intellectual development at school age. IN: VYGOTSKY, L. S.; LURIA, A. R.; LEONTIEV, A. N. **Language, development and learning.** 10ª ed. São Paulo: Ícone Editora, 2006.

VIANNA, D. M.; ARAÚJO, R. S. **Searching for Elements on the Internet for a New Pedagogical Proposal.** In: Teaching Science: Uniting Research and Practice. Carvalho, A. M. P. de (Org.). São Paulo: Thomson, 2004.

APPENDIX1: Teacher questionnaire

QUESTIONNAIRE

Training:

Time in the profession:

Series he teaches:

1. What preparation do you think teachers need in order to carry out their educational activities

using the new technologies the school has?

2. What knowledge do you believe is necessary for teachers to use new technologies at school in order to provide meaningful learning?

3. Do you believe that the use of technology can favour meaningful learning? Why?

APPENDIX2: Questionnaire applied to students

QUESTIONNAIRE

01. Which of the following factors most influence and hinder your learning of maths content?
Note: You can tick more than one option.
( ) I had no difficulties
( ) Doesn't like maths
( ) Lack of teacher training
( ) Lack of dynamic resources in the classroom
( ) Too few lessons for too much content
( ) You don't like your teacher
( ) Other factors. Which factors?

Note: From question 2 onwards, only one option per question should be ticked:
02. Do you think that using technological resources in class can help you improve your performance in the subjects you study?
( ) Yes ( ) Maybe ( ) No

03. Does the maths teacher usually use the computer lab in his or her lessons?
( ) Always ( ) Sometimes ( ) Hardly ( ) Never

04. How do you classify the purpose of using the GeoGebra 3D software?
( ) Very useful ( ) Useful ( ) Not very useful

05. How do you rate the visualisation of solids using GeoGebra 3D?
( ) Improved a lot ( ) Improved ( ) Didn't see any difference

06. What has changed in terms of learning the content you worked on using the software?
( ) Made it much easier ( ) Made it partly easier ( ) Hasn't changed at all

07. Give your opinion on the use of GeoGebra 3D software in the teaching of Spatial Geometry.

Printed by Books on Demand GmbH, Norderstedt / Germany